MW00843749

Luminos is the Open Access monograph publishing program
from UC Press. Luminos provides a framework for preserving and
reinvigorating monograph publishing for the future and increases
the reach and visibility of important scholarly work. Titles published
in the UC Press Luminos model are published with the same high
standards for selection, peer review, production, and marketing as
those in our traditional program. www.luminosoa.org

DNA, Race, and Reproduction

DNA, Race, and Reproduction

Edited by

Emily Klancher Merchant and Meaghan O'Keefe

UNIVERSITY OF CALIFORNIA PRESS

University of California Press
Oakland, California

Suggested citation: Merchant, E. K. and O'Keefe, M. (Eds.). *DNA,
Race, and Reproduction*. Oakland: University of California Press, 2025.
DOI: https://doi.org/10.1525/luminos.218

Cataloging-in-Publication data is on file at the Library of Congress.

ISBN 978-0-520-39958-7 (pbk. : alk. paper)
ISBN 978-0-520-39959-4 (ebook)

Manufactured in the United States of America

33 32 31 30 29 28 27 26 25 24
10 9 8 7 6 5 4 3 2 1

CONTENTS

ILLUSTRATIONS

FIGURES

TABLE

ACKNOWLEDGMENTS

This volume represents the efforts and support of numerous people. As the editors, we would like to thank our contributors above all, without whom this volume would not exist. We would also like to thank the Davis Humanities Institute, which funded this effort as a Transdisciplinary Research Cluster, and the DNA in Society Reading Group, out of which this project grew. We are immensely grateful to have had the opportunity over the past several years to think with our UC Davis colleagues about all of the issues represented in this book. At University of California Press, we thank our editors, Chloe Layman and Chad Attenborough (and we thank Lisa Ikemoto for connecting us with them), and the readers who provided feedback at various stages. We also thank Rajani Bhatia, Ellen Foley, and Jade Sasser for providing feedback on the introduction. Additionally:

Carlos Andrés Barragán, Sivan Yair, and James Griesemer thank Emily Merchant, Meaghan O'Keefe, and all the participants in the seminar DNA, Race, and Reproduction (which led to this volume) for creating a wonderful space to think and dialogue about the research process. At UC Davis, they are also in debt to Graham Coop (Department of Evolution and Ecology), Elihu M. Gerson (Department of Philosophy), and Brenna M. Henn (Department of Anthropology) for valuable discussion and criticisms of some of their arguments.

Mark Fedyk thanks the editors of and contributors to this volume for their many helpful suggestions and criticisms. He particularly thanks Jim Griesemer and Alok Srivastava for their additional help.

Lisa Ikemoto's chapter has been incubating since the beginning of the COVID-19 pandemic. She appreciates the invaluable feedback throughout that period from the UC Davis DNA in Society Working Group, especially editors Emily Merchant and Meaghan O'Keefe, and the excellent work of research assistants Kristi Thielen, Kurt Van de Mortel, Cailin Lechner, and Yenna Ahn.

Emily Klancher Merchant thanks Rajani Bhatia, Ellen Foley, and Jade Sasser for reading multiple drafts of her chapter. She also thanks everyone in the School of Social Science and everyone in the history of science group in the School of Historical Studies at the Institute for Advanced Study (IAS) during the 2021–22 academic year, with special thanks to Joan Scott. This chapter was written during a membership in the School of Social Science at IAS, with funding provided by IAS and by a Hellman Fellowship from UC Davis. Emily thanks Rina Bliss, Dalton Conley, and Sam Trejo for helpful intellectual comradeship during that period.

Meaghan O'Keefe thanks the Mellon Foundation. This book was, in many ways, the product of the interdisciplinary course of study in genetics and law funded by the New Directions Fellowship. She also thanks the other contributors for their feedback on drafts of the chapters. Special thanks also to Emily Merchant for all her work on this volume.

Alice B. Popejoy wishes to thank the contributors to this volume and the editors, Emily Merchant and Meaghan O'Keefe, for their thoughtful and diligent leadership on this multiyear effort, many important discussions and feedback on the ideas, and suggested revisions on an earlier draft of the chapter. Thanks also to Michael Yudell, Michelle Burton, and David Hayes-Bautista for regular check-ins that kept a steady drumbeat on the issues discussed, as well as their encouragement to continue pushing for more precise and trustworthy science. This chapter was written with support from the UC Davis Department of Public Health Sciences, acknowledging the mentorship of Brad Pollock and Rachel Whitmer. Meghan Markle and Prince Harry sharing their intimate story on Netflix was critical to the throughline narrative of this chapter, as were the stories of innumerable others whose personal experiences, courage, scholarship, and leadership have laid the foundation for this story to be told, in this way, at this time.

Tina Rulli thanks the contributors to this volume and the faculty in the UC Davis Philosophy Department for feedback on an early version of this chapter. Special thanks to Meaghan O'Keefe and Emily Merchant for extremely helpful comments and extensive engagement with the ideas in this chapter.

Introduction

DNA, Race, and Reproduction
in the Twenty-First Century

Emily Klancher Merchant and Meaghan O'Keefe

On the day we logged in, the California Cryobank, one of the country's premier sperm banks, had 242 donors available, identified by five-digit numbers. Some had also given themselves cutesy nicknames, such as "Off-the-charts smarts," "Ph.D. pianist," and "Dancing scientist." We could filter the list by various attributes of the donor: physical characteristics (height, eye color, hair color, hair texture, blood type, celebrity look-alikes), "ethnic origin" (American Indian or Alaska Native, Asian, Black or African American, Caucasian, East Indian, Hispanic or Latino, Middle Eastern or Arabic), "self-reported ancestry" (specific country or countries of origin), religion (agnostic/atheist/declined to state, Buddhist, Christian, Hindu/Sikh, Jewish, Muslim, other), "self-reported Jewish ancestry" (yes or no), and education level (bachelor, master, or postgraduate, though many donors were current students, so this could refer either to a degree they had already earned or to a degree toward which they were still working). Some donors had provided descriptions of themselves; "fun facts" listed for each included favorite animal, fantasy lunch date, and favorite food. The price of sperm depended on how much personal information a donor was willing to share with potential offspring. Sperm from donors who agreed to disclose their identity fetched a premium, suggesting an expectation among prospective parents that their children will want to know where their DNA comes from.

This imagined desire stems, at least in part, from scientific and popular understandings of which characteristics genes carry from one generation to the next. Many of these assumptions were on display in the California Cryobank's database. Genetics reaches into numerous domains of human life in the twenty-first century, but it is in the crucible of reproduction—whether accidental or planned, technologically assisted or old-fashioned—that individuals, couples, and families are forced to confront the science and mythology of genetics and make decisions about which characteristics of themselves and each other they want to reproduce. These decisions are most obvious in the realm of assisted reproductive technology, where the fertility industry represents them as choices available to paying (and typically white) customers. But people who get pregnant from sex also come face-to-face with ideas about genetics when health-care providers offer—or even push—prenatal testing options ranging from carrier screening to amniocentesis, though the availability of these tests and even the opportunity to opt out of them depends on the pregnant person's access to health care. In generating new life, or simply in contemplating reproduction, genetics inevitably comes to the fore via ideas about what makes individuals distinct from one another, what generates affinity between family members, and what endows people with social value. In those ideas, popular and scientific conceptions of race and the patterning of human difference are never far from the surface.

This book emerged from a series of conversations among UC Davis faculty members working across the humanities, social sciences, and natural sciences, as well as the School of Law, the School of Medicine, and the School of Nursing, about how popular, religious, scientific, legal, and medical understandings of genetics come together in a variety of settings to profoundly shape the contemporary human experience. Some of the essays contained here reflect interdisciplinary collaboration; the others were written by individual scholars trying to reach audiences outside of our own disciplines and, we hope, to influence conversations about genetics in a variety of settings, from laboratories to doctors' offices, courtrooms, schools, and houses of worship. All of the essays grapple with how popular and professional understandings of DNA influence society. Some center race, others center reproduction, and others center the intersection between race and reproduction. It is our hope that this volume will help a broad variety of readers think critically about all of the ways they confront DNA and ideas surrounding DNA in their daily lives.

All of the locations listed above—laboratories, doctors' offices, courtrooms, schools, and houses of worship—are places where people with different kinds and levels of expertise interact. They are also sites where significant institutional power differentials are enacted and reinforced by the discourse structures that form knowledge systems. By design, terms of art, jargon, and other habits of language used within disciplines and professions can make it difficult for nonspecialists to engage meaningfully with ideas. In this volume, we have tried to steer away from insider talk and to explain some of the more technical topics from our respective disciplines in everyday language. This, of course, means that some of what we say will be new to all readers, but parts may also be extremely familiar to some. We hope that our work will be broadly useful for those who have spent many years studying these topics as well as those who are just beginning their studies and those who are simply curious.

In this volume, we treat reproduction as both a site of inquiry and an analytic tool with which to interrogate the human attributes that are widely believed to inhere in DNA. Our contributors focus on race as a social category that points to biology for justification. Race gets perpetuated in part through decisions people make about reproduction, but also gets reproduced through inequitable access to reproductive (and other) health care, along with numerous other social, political, and economic goods. The contexts in which people make reproductive choices are structured by a variety of historical and social circumstances, including those that produce and perpetuate racial difference. While not all chapters directly address reproduction, in one way or another, each addresses the complex, confusing, and inconsistent ideas about the relationship between genetics and identity that circulate among the American public. Understanding the historical, legal, and scientific construction of race and ethnicity is integral to understanding how reproductive choices are, in a sense, prestructured to both enforce and perpetuate notions of genetic identity, race, and ethnicity.

Although some chapters have a wider geographical reach, we focus mainly on the United States for three reasons. First, for most contributors, the United States is the social, historical, legal, medical, political, and religious landscape we know best. Second, race is a historically, socially, politically, and legally constructed set of categories that necessarily differ from place to place and time to time. Nonetheless, given the global hegemony of the United States, the racial categories used here get exported to other parts of the world as well. Third, the United States has fewer regulations surrounding assisted reproductive technologies and gamete donation than most other countries, making it a global center of reproductive tourism. In 2019, 2.1 percent of all births in the United States were the result of assisted reproductive technologies. In California—a major hub for reproductive tourism due to lax regulation—rates range from 3.5 to 5.5 percent of births.[1] The United States is therefore ground zero for debates about how new genetic technologies can and should become part of reproductive decision-making.

We have grouped the chapters in this book into three parts—"DNA and Race," "DNA and Reproduction," and "Race and Reproduction"—though most chapters address all three of our key themes in one way or another. The first part, "DNA and Race," explains how older ideas about race have shaped and been reshaped by new genomic technologies. The second part, "DNA and Reproduction," examines how these new genomic technologies have become part of the landscape of fertility medicine. The third part, "Race and Reproduction," explores how genetic understandings of race and family (including the human family) influence one another. The remainder of this introduction provides some context for each of the three parts of the book and indicates how later chapters will further elucidate specific themes.

DNA AND RACE

Race and ethnicity, though lacking any stable meaning or clear boundaries, have formed the central axes of identity in the New World since the arrival of European colonists and enslaved Africans. In many ways, race and ethnicity are forms of what the anthropologist Jonathan Boyarin has described as "the rationalization and regulation of identity and difference" that long anchored Christian anti-Semitism.[2] Early efforts to distinguish Christianity from Judaism involved concepts of bodily difference that later reappeared in race science.[3] Exactly what those bodily differences are, however, has continually changed as social and political circumstances have required different criteria of inclusion and exclusion, and as science has continually failed to identify any clear lines of demarcation between racial or ethnic groups, or even between the concepts of race and ethnicity. Indeed, the groupings themselves have continually changed, with no expert agreement on either how many races and ethnicities there may be or on whether any given identity descriptor refers to a race, an ethnicity, or something else altogether.

Race is typically deployed as a classificatory schema that makes reference to the large-scale (often continental) geographic origin of our ancestors. It is thought to be perpetuated by reproduction within categories and interrupted by reproduction between categories. Race categories differ from place to place and have changed over time.[4] Currently, the US government recognizes five races: White, Black, Native American or Alaska Native, Asian, and Pacific Islander or Native Hawaiian. Since 1995 it has allowed individuals to identify with more than one category.[5]

Ethnicity is a separate set of social categories that may be subracial (for example, referring to nationalities within continents) or may cut across racial divisions (for example, referring to language, culture, and/or religion). The US government recognizes only two mutually exclusive ethnicities—Hispanic and non-Hispanic—and allows people of either ethnicity to identify with any number of races (but the US Census requires people of both ethnicities to identify with at least one race).[6] The fact that the US government's definition of ethnicity differs so dramatically from colloquial uses of the term indicates the social and political constructedness and the geographical, cultural, and chronological contingency of the entire concept.

By the middle of the twentieth century, when skin color, hair texture, skull size, and face shape had all failed to produce clear boundaries between presumed racial categories, social scientists concluded that race was socially constructed. Natural scientists, on the other hand, sought a biological basis for it in the emerging science of population genetics. In 1950 the UN Educational, Scientific and Cultural

Organization (UNESCO) published a statement declaring that "the species *homo sapiens* is made up of a number of populations, each one of which differs from the others in the frequency of one or more genes," and that "a race, from the biological standpoint, may therefore be defined as one of the group of populations constituting the species *homo sapiens*."[7] But even that definition didn't adequately fit colloquial notions of race. In 1972 the population geneticist Richard Lewontin found that the majority of human genetic variation occurs *within* racially defined groups rather than between them.[8] In 1977 the US government established a set of race categories for statistical purposes, explicitly stating that "these classifications should not be interpreted as being scientific or anthropological in nature."[9] Nonetheless, some scientists continued to pursue a genetic basis for these categories.

The first two decades of the twenty-first century in particular (since the completion of the Human Genome Project) have seen what critical science scholars Barbara Koenig, Sandra Soo-Jin Lee, and Sarah Richardson describe as "a vigorous reassertion of the coupling of race and genes."[10] Population geneticists have made strenuous efforts to link racial categories to genetic differences, convincing large swaths of the public without producing empirical evidence that race categories have any genetic basis.[11] The overwhelming preponderance of research has found that genes and physical characteristics vary clinally—that is, gradually and continuously—across space, and that genes corresponding to physical characteristics that are commonly thought to cluster in racial groups (for example, skin color and hair texture) vary independently of one another.[12] Large-scale racial definitions therefore fail to capture actual human variation, which is—as the anthropologist Jonathan Marks observes—"historically ephemeral," "genetically porous," and "culturally bounded."[13]

These new efforts to find a genetic basis for race often use the language of *genetic ancestry* to avoid charges of racism.[14] In its most technical sense, genetic ancestry refers to the genealogy of each of our genomic loci. At every point on our genomes, we have inherited one allele from our mother and one from our father. The alleles we inherited from our mother could have come either from her mother or from her father (same for the alleles we inherited from our father), and before that from those people's mothers or from their fathers, and so on. Genetic ancestry describes this path of genetic inheritance.[15] Yet the term has come to mean something else: the race or ethnicity of the people in your genetic ancestry. Notice this slippage. According to one definition, our genetic ancestry consists of the *people* from whom we have inherited DNA. According to the other, our genetic ancestry consists of the social categories those people would have identified with or been classified into, were they alive today. As a result of this slippage, "genetic ancestry" has become a seemingly scientific substitute for race used by scientists and nonscientists alike.

There is a further slippage, which is that a DNA test can't tell you either who your ancestors were (unless their DNA is available for comparison) or to which social categories they belonged (or with which they would currently identify). Genetic ancestry (as generally understood) is instead determined by identifying genomic similarities between the person in question and contemporary reference samples from various parts of the world. By triangulating in this way, a test might identify someone as having x percent "African genetic ancestry" or y percent "European genetic ancestry" or z percent "Asian genetic ancestry," when what they actually identify are percentages of a person's genome that *resemble those of people currently living in Africa, Europe, and Asia.* These geographical designations are salient to Americans only because they map onto our continentally based racial categories.[16] The concept of genetic ancestry therefore suggests that our ancestors belonged to discrete, genetically bounded populations that correspond to present-day notions of race and ethnicity, and that these identities can be read in our DNA, even though such groupings have never actually existed.[17]

Capitalizing on this line of research, numerous companies now sell Americans quantitative assessments of their "genetic ancestry," equating genetic similarity to reference samples in various parts of the world with an imagined biological autochthony in those parts of the world.[18] Indeed, members of white nationalist groups sometimes use the results of such tests as a basis for membership.[19] These products are designed by scientists who move readily between academia and industry, shuttling scientific and popular notions of ancestry back and forth and blurring them together as they travel.

In the chapter "Are People Like Metals? Essences, Identity, and Certain Sciences of Human Nature," the philosopher Mark Fedyk explores a tension between the statistical logic of classification based on ancestry testing, which partitions a person's unitary identity among categories that sum to 100, and the logic of people's "real-world" identities, which are often multiple, intersecting, and overlapping. For many Americans, DNA ancestry tests—whether people take them or not—play a role in the development or validation of racial or ethnic identity and endow those identity categories with a veneer of biological reality and scientific authority.[20] Fedyk argues that these identity categories, largely invented by the companies that sell the tests, have come to be understood, by scientists and the public alike, as "real essences" that somehow explain who we are and why we are the way we are, even when they don't resemble the ways in which people actually form and express their social identities. Ultimately, he insists that genetic models of identity are no more accurate or scientific, and may be less informative, than understandings of identity that come from the social sciences and humanities.

Although racial categories are fundamentally and irreducibly *social* taxonomies, they are nonetheless *real*. Social classification has material effects that produce biological differences, most visibly in health disparities.[21] The historian Terrence Keel argues that—contrary to the popular notion that science is opposed to religion—Christian thought about race and identity has been integral to both historical and contemporary scientific accounts of human diversity. He suggests that "racial reasoning strategies" rooted in Christian intellectual history help explain current researchers' preference for "nature" as the cause of different health outcomes rather than social structures that give rise to disparities.[22] The concern over differentiating Christian Europe from "others" and present inclinations toward supposedly natural explanations tend to give precedence to fixed racial categories. The widespread perceptions among scientists and biomedical researchers that the causes of disease are primarily genetic rather than environmental and that racial health disparities are rooted in genetic difference has, perhaps paradoxically, led to racial exclusivity in genomic research. The vast majority of genomic research today is done on white-identified subjects, and white people are its primary beneficiaries.

But how can we argue for greater diversity in genomic research without reinforcing the mistaken and racist idea that race is a genetic category? This is the question the philosopher Tina Rulli answers in her chapter, "A Colorful Explanation: Promoting Genomic Research Diversity Is Compatible with Racial Social Constructionism." Calls to diversify genomic research often rely on and reinforce the assumption that humans are members of discrete "populations" or "ancestry groups," substituting these scientific terms for race and ethnicity.[23] Rulli begins from the scientific findings that human genetic diversity is geographically patterned and that it does not cluster into groups, much less groups that map onto US racial categories.[24] She argues that, due to the history of the ways in which racial categories have been socially, legally, politically, and economically constructed, two things are true. First, a database or study that includes only or primarily white-identified people is also highly likely to lack genetic diversity. Second, individuals who are differently racialized occupy different social, economic, and physical environments, and the same genes could behave differently in these different environments. Rulli contends that, while race is unlikely the *best* proxy for genetic diversity in research settings, it is possible to use race as a proxy for genetic diversity without endorsing racial essentialism or race realism. She cautions, however, against using an individual's racial identity as a proxy for their own genotype in clinical or other settings. More racial diversity in a research sample will likely increase the genetic diversity of the sample, but not because specific racial categories necessarily reflect specific genotypes.

The 1950 UNESCO Statement on Race equated racial groups with populations. However, populations are heuristics that might be useful for answering certain research questions, not biological realities.[25] Similarly, while some researchers treat "genetic ancestry" as a set of fixed, naturally defined categories, the chronological and geographical scale at which populations are identified is totally arbitrary and depends on the question being asked. For example, the same researchers might define populations in terms of countries or towns for one project and in terms of continents for another, and a group of people that might be classified as members of one population at a given point in time (for a specific research purpose) might be classified as members of two separate populations at an earlier or a later point (or for a different research project). For the purposes of medical and social scientific research in the United States, however, genetic ancestry is usually identified at the continental scale, perpetuating the illusion that the social concept of race is built on a biological substrate of population divisions that have not changed from time immemorial.[26] As the bioethicist Jonathan Kahn explains, "the idea that there are somehow 'pure' types of African, European, or Asian DNA is a fiction, constructed not only by artificially bounding geographic areas but also by arbitrarily designating distinct points in time as marking the temporal moment of purity."[27]

This myth of a "temporal moment of purity," when humans fit neatly into discrete categories, is bolstered by the concept of *admixture*, a term that originated in race science to refer to interracial reproduction and now refers to the mixing (by reproduction) of two populations understood to have been separated in space and/or time, such that both lineages are identifiable (relative to some kind of reference) in the allele frequencies of the offspring. Since admixture refers to populations, and populations are local and relative rather than universal and absolute, admixture, too, is a local and relative concept that can be scientifically useful but doesn't identify anything in the real world. Like genetic ancestry, however, the concept of admixture can reify existing ideas of what constitutes populations and, by extension, racial and ethnic categories. Also like genetic ancestry, the concept of admixture can have social consequences within and beyond scientific contexts.

In their chapter, "Eventualizing Human Diversity Dynamics: Admixture Modeling through Time and Space," the anthropologist Carlos Andrés Barragán, population geneticist Sivan Yair, and philosopher James Griesemer show how the concept of admixture is used in the modeling of ancient migrations, specifically concentrating on the peopling of the Americas. They track the term as it has emerged and proliferated in the scientific literature and show how this genomic knowledge has made its way into popular contexts and back into science. The slipperiness of the term and the opportunities for misunderstanding make admixture a particularly valuable case study. For example, what counts as distinct populations and what timescale marks the divide between introgression, admixture, and migration are still unsettled. Given the ramifications of dividing and defining populations, Barragán, Yair, and Griesemer suggest strategies for reducing misunderstandings of admixture modeling between scientists and those outside the scientific community. They end with a consideration of both the limits and the potential of modeling.

The three chapters in "DNA and Race" consider how recent genomic research has reconfigured and thereby reinforced much older ideas of human difference. There is, perhaps, no better place to see the materialization of popular understandings—held by the public and medical professionals alike—than in the fertility clinic, where individuals and couples make decisions that are thought to have bearing on the racial identity and social characteristics of their future children. The next part of the book turns to assisted reproductive technology to see how Americans understand race and other aspects of individual identity—specifically intelligence—to inhere in our DNA.

DNA AND REPRODUCTION

Reproduction typically combines the DNA of two individuals, though it is now possible to add the mitochondrial DNA of a third. Nowhere is the idea of building a baby from component parts starker than in the world of gamete donation, where would-be parents choose the person from whom half of their child's DNA will come. Despite the language of "donation," gametes are in fact bought and sold.[28] Most people using donor sperm purchase it from a commercial sperm bank, such as the California Cryobank. People using donor eggs can purchase them from an egg bank, such as Santa Monica Fertility, but given the fact that fresh eggs are somewhat more likely than frozen eggs to produce a live birth,[29] many egg recipients contract with individual donors, either directly or through an agency such as Circle Surrogacy, which maintains a database of individuals interested in selling their eggs. Donor sperm can be used in either intrauterine insemination or in vitro fertilization (IVF); donor eggs must be used in conjunction with IVF. Given the low success rates of IVF, the marketing of gametes is as much the selling of hope or the satisfaction of having exhausted all avenues for remedying infertility as it is the selling of fertility itself. In this selection process, sperm banks and egg brokerages invite prospective parents to consider the process of amalgamating their own DNA with that of a donor, or to consider how the characteristics of two donors will complement one another. In this amalgamation process, clinics and customers typically focus most heavily on the donor's race and/or ancestry and on their intelligence and/or educational attainment.

Direct-to-consumer (DTC) genetic testing companies present customers with a fractionated identity: parts that add up to a whole. One long-running ad for 23andMe showed a racially ambiguous woman traveling the world. As she moves from place to place, viewers realize that she is exploring in the world the ancestries she "discovered" through genetic testing: 29 percent East Asian, 3 percent Scandinavian, 46 percent West African. In each place, she seems to fit right in with the locals, presumably because she shares something fundamental with them—a genetic identity—even though they have lived very different lives. When prospective parents choose a gamete donor, that person's race or ancestry is never left up to chance. Sperm banks typically use several different metrics for this. As we saw in our perusal of the California Cryobank catalogue, donors are asked for their "ethnic origin," which elicits US Census race categories; their "self-reported ancestry," which elicits finer-grained identities, usually corresponding to countries; and whether or not they have "Jewish ancestry," a concept that blends religion, race, and ethnicity. Customers are thereby forced to consider the identity of their donor in these terms, and some may even imagine what they would want their future child to learn if they were to send a vial of saliva to 23andMe.

In the past, clinics typically abided by the "one-drop" rule of hypodescent for mixed-race donors—meaning that donors with multiple identities were classified according to the one with the lowest social value—to make certain that characteristically minority phenotypes do not "surprise" white consumers.[30] More recently, however, clinics have allowed donors to identify with more than one race. Some (including the California Cryobank) make available the kinds of ancestry percentage breakdowns offered by DTC genetic tests for donors who choose to undergo ancestry testing. These results are reported in geographical terms, but color-coded such that all locations within a given continent are different shades of the same color. The report thus allows customers to easily translate between ancestry and race, implying that race is not just quantifiable but also precisely measurable.

In the chapter "Selling Racial Purity in Direct-to-Consumer Genetic Testing and Fertility Markets," the legal scholar Lisa C. Ikemoto argues that the quantification of donor ancestry, and its conflation with race, advances a dangerously incorrect model of racial diversity. The model implicitly assumes that there are— or once were—"pure" races, and that genetic ancestry testing reveals how these ur-races have combined in individual bodies. Although DTC ancestry testing companies and sperm banks that use ancestry testing appear to celebrate diversity, this model actually naturalizes racial inequality as the product of separate evolutionary processes. As Ikemoto points out, the idea of racial purity emerged in the context of global white supremacy and generally serves to protect the exclusivity of white privilege. In the realm of gamete donation, the use of ancestry testing as a marketing tool invites consumers to literally curate the racial identity of children and families, down to the percentage point.

In addition to classifying donors on the basis of several dimensions of biogeographical identity, sperm banks and egg brokerages also tout the intelligence and educational success of their donors. To be sure, they are responding to market demand, which is driven by a widespread belief, originating in the eighteenth-century eugenic thought of Francis Galton, that a person's socioeconomic status is determined primarily by their intelligence and that intelligence is determined primarily by biological heredity. The consumer-choice approach to gamete donation emerged with the rise of for-profit cryobanks in the 1970s.[31] Alongside these was the Repository for Germinal Choice, a sperm bank established in 1980 by the optometrist and businessman Robert Klark Graham. An avowed eugenicist, Graham sought to make the sperm of Nobel Prize–winning scientists available (for free) to high-IQ women in an effort to stem what he saw as the genetic deterioration of the US population.[32] Prior to the 1980s, when the HIV/AIDS epidemic spurred the rise of human sperm freezing, most sperm was used fresh, obtained from the donor on call—or the one who had most recently made a donation— at the time the recipient came in for her appointment.[33] Donors were typically recruited from universities and selected by doctors (not recipients) on the basis of their looks and intelligence. DNA testing has recently revealed that some fertility doctors also impregnated unwitting patients with their own sperm.[34] The use of donor sperm in intrauterine insemination was controversial and legally question-able for a long time, and often embarrassing for infertile husbands even after the legality was settled.[35] To provide couples with plausible deniability, fertility clinics sometimes mixed donor sperm with the sperm of the patient's husband.[36] Until fertility treatment became a big business around the turn of the twenty-first cen-tury, doctors served as the gatekeepers to fertility treatment, deciding who was worthy of receiving donor sperm and other interventions.[37] Most fertility doc-tors restricted treatment to white women married to white men; some openly acknowledged their eugenic aims.[38]

Egg donation became possible more recently than sperm donation. For people trying to make babies, purchasing eggs is more expensive than purchasing sperm because retrieving eggs is more physically invasive, time-consuming, and risky for the donor. The technology for successfully freezing and thawing eggs became avail-able much later than the technology for successfully freezing and thawing sperm. Sperm donors are typically compensated somewhere in the range of $4,000 for a series of weekly or twice-weekly donations over a period of several months.[39] Pur-chasing a vial of sperm will run you approximately $950 to $1,150. For eggs, there is less of a gap between what a donor makes and what a recipient pays, though the recipient usually also pays the donor's medical expenses, and the process of getting pregnant with donor eggs is typically more complicated and costly than is the process of getting pregnant with donor sperm. It is difficult to determine how much eggs cost on average because so many transactions are conducted privately. Additionally, egg donation agencies often let donors set their own compensation, which contributes to price variation. *Wired* has estimated that donors typically make between $8,000 and $10,000 per cycle, but they can charge up to $50,000 or more if they have desirable traits, including higher levels of educational attain-ment or matriculation at fancier universities.[40]

Once gamete donors are selected, or if a couple uses their own gametes, would-be parents need to decide which embryos to carry to term. This choice is most evident in the case of IVF, where patients often produce more viable embryos than they want to implant. But even people who get pregnant through sex or intrauter-ine insemination need to make choices about whether to undergo genetic screen-ing or testing that could influence their decision about whether to continue the pregnancy. Indeed, the very existence of such tests is premised on the idea that certain results would lead to a decision to terminate.[41]

Until only a few years ago, in utero genetic testing (through amniocentesis or chorionic villus sampling) and preimplantation genetic diagnosis (in conjunc-tion with IVF) were used only to identify chromosomal anomalies (such as aneu-ploidy) or straightforwardly genetic conditions that were known to run in parents' families, such as cystic fibrosis or sickle cell disease. By "straightforwardly genetic conditions," we mean diseases or other medical conditions that are caused by identifiable genomic variants and where the biochemical mechanism by which the variants cause the disease is more or less understood. Many parents faced with the prospect of having a child with a serious genetic disease will choose to terminate a pregnancy or discard IVF embryos that carry the variants responsible, particularly for diseases such as Tay-Sachs. Children with Tay-Sachs suffer seizures, vision and hearing loss, and paralysis, and generally live to only four or five years old. Other conditions that are not fatal but may result in disability present ethical quandaries. Disability activists have expressed serious concern at the prospect of disability screening, arguing that the medicalization of disability results in a perception that "disability invariably equals tragedy," an idea at odds with the lived experience of many people with disabilities.[42] In addition to the medical community's attitude toward disabilities, many parents making reproductive decisions about having children with disabilities are often not disabled themselves and so may have difficulty understanding or anticipating the experiences of people with disabilities.

Technologies for the genetic testing of embryos (prior to implantation) or fetuses (in utero) are often presented as tools for making "healthy" children. But the definition of "healthy" children has become more capacious with the development of new screening technologies. Since the completion of the Human Genome Project, medical geneticists have developed new tools to identify genetic predispositions for conditions that are not straightforwardly genetic, such as heart disease, diabetes, and schizophrenia, which are believed to run in families but are not caused by a single gene and for which the biochemical mechanisms of causation are not known. A relatively new approach for identifying the "genetic architecture" of such complex diseases is the genome-wide association study (GWAS), which tests millions of loci across the genome for single nucleotide polymorphisms (SNPs—variations in individual nucleotides, the components of DNA) that correlate with the disease in question. Unlike straightforwardly genetic diseases like Tay-Sachs or Huntington's, where the biochemical function of the variant is known and the test can reliably predict the current or future presence of the disease, GWAS show that people with some constellation of variants may have some propensity to develop a disease predicated on environmental factors that may or may not be known. The result is a formula for calculating an individual's polygenic score or index, which is widely (but often incorrectly) interpreted as their genetic propensity for developing the given condition.

Scientists and other observers have criticized polygenic scores for a number of reasons. Since the conditions that are subjected to GWAS are heavily influenced by such nongenetic factors as diet, smoking, stress, socioeconomic status, and exercise, the polygenic scores produced by GWAS provide limited utility in predicting disease. Even if such risk factors could be accounted for, polygenic scores predict disease far better among white-identified people than among people of color, because the vast majority of GWAS include only white-identified people in their discovery samples. The use of polygenic scores in medical settings (for example, to guide treatment or screening plans) therefore threatens to increase health disparities between white patients and patients of color.[43]

Racially structured differences in the predictive power of polygenic scores stem from the racial structure of GWAS themselves. Potential GWAS participants are classified by the continent(s) represented in their "genetic ancestry," and typically only those with continentally homogeneous ancestry are included. Due to the need for enormous samples and the fact that most GWAS are done by researchers based in the United States or Europe, individuals with exclusively "European genetic ancestry" are massively overrepresented in GWAS.[44] As of November 2023, the GWAS Diversity Monitor showed that GWAS participants were (in terms of "genetic ancestry") 94.7 percent European, 3.56 percent Asian, 0.18 percent African, 0.49 percent African American or Afro-Caribbean, 0.33 percent Hispanic or Latin American, and 0.68 percent other.[45] However, recent research has demonstrated that there is enough diversity *within* continents to undermine the findings of GWAS in European discovery samples that were previously thought to be relatively homogeneous.[46]

According to the prevailing "out of Africa" model of human history, individuals with exclusively "European genetic ancestry" comprise only a small fraction of the world's genetic diversity. In contrast, people with more recent "African genetic ancestry" encompass a great deal more diversity.[47] The construction of genetic ancestry for the purpose of GWAS closely matches the construction of race in the United States: GWAS typically include people with *only* "European genetic ancestry," just as the white identity category has historically been constructed to include individuals with *only* European ancestors. People who do not identify as white are therefore more likely to have genetic variants that have not been studied and are thus less likely to benefit from existing genomic research. At this point, polygenic scores for individuals of recent African descent are often no better than random chance for predicting disease risk.[48]

In spite of the fact that polygenic scores don't do a great job of predicting disease even among white-identified people, several new companies have begun to make polygenic embryo screening available to couples and individuals undergoing IVF. In the United States, Genomic Prediction and Orchid use polygenic screening to estimate risk for a number of diseases and medical conditions—including breast cancer, prostate cancer, diabetes of both types, coronary artery disease, and schizophrenia—to prioritize embryos for implantation. The price tag is in the thousands of dollars.[49] Only Genomic Prediction, whose motto is "choice over chance," acknowledges on its website that polygenic scores "perform less well when applied to individuals from distant [from European] ancestry groups (e.g., African ancestry, East Asian ancestry)." In the age of GWAS, "healthy" has come to mean not just disease free, and not just free of genes that are known to cause disease, but also as free as possible of disease risk.[50]

When Genomic Prediction first offered polygenic embryo screening in 2019, its scope exceeded disease risk: embryos were also tested for the risk of short stature and "intellectual disability," the company's disingenuous label for low predicted educational attainment.[51] This latter test was made possible by a series of GWAS of educational attainment that occurred over the past ten years.[52] Due to public distaste, Genomic Prediction quietly dropped "intellectual disability" from its menu of tests at the end of 2020. A 2021 article in the *New England Journal of Medicine* explained that selecting embryos on the basis of the polygenic score for education is unlikely to have the desired effect on children's intelligence or education levels.[53] Few of the SNPs that *predict* high educational attainment can be said to *cause* high educational attainment in any meaningful way. Most simply correlate with environmental predictors of high educational attainment, such as having well-educated parents and living in wealthy neighborhoods.

The chapter by the historian Emily Klancher Merchant, "Reproducing Intelligence: Eugenics and Behavior Genetics Past and Present," places the GWAS for educational attainment into a historical trajectory that reaches back through the behavior genetics of the mid-twentieth century to the eugenics of the late nineteenth century. Merchant demonstrates that eugenics inspired early twentieth-century efforts to measure intelligence and its heritability, or the amount of variance in a sample that is due to genetic variation rather than nongenetic variation, which formed the foundation for classical behavior genetics. GWAS for educational attainment provide a molecular update to this eugenic research agenda, but have not improved scientific understanding of how genes might contribute to individual differences in intelligence, educational attainment, or socioeconomic status. Instead, research probing the results of these GWAS has undermined the eugenic claims that inspired the field, demonstrating that, if genes contribute to these differences at all, direct genetic effects are very small and are largely overwhelmed by nongenetic factors, primarily childhood socioeconomic status.[54] Yet, as science turns up more and more evidence that the effects of genetics on individual differences in intelligence, educational attainment, or socioeconomic status are indeterminate at best, the scientists who produce these results increasingly publish books and articles for the general public claiming that DNA plays a decisive role in these matters.[55]

These books further popular but incorrect ideas that intelligence and socioeconomic success are genetically determined,[56] and these ideas are reflected in and perpetuated by the landscape of gamete donation. California Cryobank has locations in Los Angeles, Cambridge, New York, and Los Altos, which recruit donors from prestigious nearby universities, including USC, UCLA, Harvard, MIT, NYU, Columbia, Stanford, and UC Berkeley. Egg donors are also often recruited from universities and can request higher compensation if they have more education. DonorNexus, an egg brokerage, has a starting charge of $32,000 for its "premier egg donor" program, which allows prospective parents to select donors "with a specific set of desirable traits, such as higher education, rare ethnicities, professional athletes, musicians, or models."[57] Seeking to highlight donors' youth as well as their accomplishments, the agency describes them as "smart and ambitious *young* women . . . *in the early stages of establishing themselves* in respectable lines of work" (emphasis added). Recognizing that egg purchasers value both beauty and brains, DonorNexus touts its premier donors as "fashion models, beauty pageant queens, actresses, tv hosts, social media influencers" on the one hand and as having "accomplished impressive academic milestones, such as engineering degrees, various graduate degrees, high SAT and ACT scores, law degrees, medical degrees, and PHD candidates [*sic*]" on the other. Such marketing indicates the widespread belief that intelligence, like appearance, is strongly rooted in DNA and therefore is transmitted by our gametes.

Research in molecular behavior genetics has not been limited to GWAS of educational attainment. Following the 2017 release of data from the UK Biobank, a flurry of GWAS claimed to identify the "genetic architecture" of just about every imaginable behavior or social outcome.[58] One of the most controversial was a GWAS of same-sex sexual activity.[59] The suggestion that sexual orientation could have a genetic component is not new. In 1993 the geneticist Dean Hamer and his colleagues found apparent evidence that male homosexuality correlated with certain markers on the X chromosome.[60] The idea that sexuality was genetic seemed liberatory to many LGBTQ Americans, indicating that sexuality was inborn rather than a matter of individual choice or pathology.[61] To others, however, the possibility of a "gay gene" raised the specter of eugenics: if sexuality were genetic, then nonheteronormative sexual orientations could be selected against. To still others, the idea that LGBTQ identity was acceptable only because people "couldn't help it" was both condescending and constraining, especially for people who identified as bisexual. It's also worth pointing out that, while most conservative Protestants believed that you could "pray the gay away," the Catholic Church was willing to accept that sexual orientation may be innate. That does not mean, however, that the Church condoned homosexual activity; it instead taught that such people must remain chaste.

Prior to the turn of the twenty-first century, most American doctors felt that, regardless of whether sexuality was determined by genetic or environmental factors (or both), LGBTQ individuals or couples should not have children. As a result, single people and same-sex couples were denied access to assisted reproductive technologies by the doctors who controlled them.[62] In many countries where assisted reproductive technologies are more heavily regulated than in the United States, these services are still limited to heterosexual couples. Even in the United States, the Food and Drug Administration bars commercial sperm donation by men who have had sex with men in the five years prior to donating. While this restriction is ostensibly intended to protect recipients from HIV infection, there is actually little risk of HIV infection from commercially available sperm, as donors are required to be tested for HIV and samples are quarantined for six months, at which point donors are retested.

Between 1993 and 2019, research on supposed "gay genes" remained inconclusive, neither validating nor invalidating theories about the heritability of sexual orientation. The 2019 GWAS of same-sex sexual activity suggested that loci across the genome influence sexuality.[63] For those who worked on the study, its results appeared to demonstrate the naturalness of same-sex sexuality without identifying one or two genes that could be selected against.[64] It nonetheless raised considerable concern among LGBTQ geneticists affiliated with the Broad Institute, where the study was carried out.[65] These concerns were vindicated when an app titled "How Gay Are You?"—purporting to calculate an individual's polygenic score for "gayness" using the results of the 2019 study—appeared on the app store GenePlaza.[66] The lead scientists on the 2019 study responded with an open letter denouncing the app and claiming that its developer had misappropriated the results of the study, which, the authors claimed, were not to be used for individual prediction. But individual prediction is exactly how polygenic scores are used in medical genomics and by Genomic Prediction and other companies offering polygenic embryo screening.

Today, parents undergoing IVF can choose to receive a "report card" for each embryo, indicating its risk level for a variety of complex diseases. They can also download raw data for each embryo and then upload the data to any of a variety of websites offering to calculate polygenic scores for educational attainment. Scientists expect that it will not be long before embryo "report cards" include predictions of each embryo's future IQ, height, sexuality, and aptitude for particular vocations.[67] Recent studies have indicated that prospective parents would welcome the ability to select embryos on the basis of such information.[68] Already, parents using gamete donation have the ability to carefully curate the racial composition of their families. The chapters in "DNA and Reproduction" examine the racist and eugenic motives behind these opportunities and consider their potential consequences.

RACE AND REPRODUCTION

As a social category, race has always depended on social institutions—particularly families—to perpetuate it. Reproduction is fundamentally a technology for making families, and the final section of the book turns to the racialization of families and children within them. As we have noted, popular ideas about the genetic foundation of race, ethnicity, and other social characteristics typically make their way into reproductive decision-making when individuals or couples decide to use donor gametes. Whether one goes to a sperm or egg bank, or to an egg broker, what is available is notionally driven by "client choice." In theory, clients can decide exactly what they want in a vast "genetic supermarket," which—in the most optimistic version—is imagined as having "the great virtue" of "involv[ing] no centralized decision fixing the futures of human type(s)."[69] In actuality, gamete banks and donation agencies seek out the donors who prove the most marketable, and what is marketable closely tracks existing hierarchies.[70] Indeed, while some donor characteristics, such as occupation, are presented to prospective parents as options, others, such as disability, are so stigmatized as to be excluded from the outset.

Genetic ancestry maps present nationalities or ethnicities as finer divisions of continental racial groups. The dynamics of the fertility market, however, suggest that consumers are more concerned about the race of donors than about their ethnicity or nationality. For example, white prospective parents from Western Europe and North America frequently cross national borders (or fly donors across borders) to procure "white" eggs from countries such as Ukraine, South Africa, or the Czech Republic, where donors typically receive less compensation than donors in Western Europe or North America.[71] Gamete recipients seem to care more about the ethnicity of donors when that ethnic identity also has a religious dimension. The medical anthropologist Daisy Deomampo describes an egg purchaser deliberately choosing a Hindu donor because he seemed to "believe that religious identity was . . . embedded in genetic ties."[72] Along similar lines, in addition to filtering potential donors by religion, California Cryobank allows customers to filter by "Jewish ancestry," suggesting that—on some level and for some religions only—a person's ancestors' beliefs and practices are encoded in their DNA. Definitions of race and ethnicity, and the categories used to distinguish people along these axes, are conventions, not facts of nature.[73]

Sperm and egg purchasing sites indicate that brokers and recipients view race and ethnicity as both nonnegotiable and reducible to searchable categories, while other traits, such as education, occupation, and special abilities, may be opportunities for negotiation. Sperm banks and egg brokerages typically present race as a category of consumer choice, even as they rigorously enforce racial boundaries. Staff often take it upon themselves to match donors and recipients on the basis of phenotype and/or identity,[74] or to restrict the use of gametes from white donors to white recipients.[75]

The idea that race and ethnicity are transmitted genetically stems from and supports the institutions that ensure they are transmitted from generation to generation socially and legally. In the purchase of sperm, race is so important to customers that many sperm banks color-code vials according to the race of the donor in order to avoid the kind of mix-up that led Jennifer Cramblett, who is white, to sue Midwest Sperm Bank after the birth of a daughter who was conceived with the sperm of a Black donor instead of the white donor Cramblett had selected.[76] This case indicates two things: First, that whiteness is a kind of property that children inherit from their parents. As the legal scholar Patricia Williams explains, Cramblett's "claim was explicitly based on the deprivation of whiteness as a trait she thought she was purchasing."[77] The sperm mix-up prevented Cramblett's child from inheriting her racial status. In other words, it prevented her from bequeathing her white privilege to her child and denied her what she considered a "legal right to a monoracial family."[78] Second, and perhaps more obviously, this case indicates that race is widely believed to be transmitted through sperm and eggs to the exclusion of other biological or social mechanisms. For this reason, white would-be parents are typically more likely to choose a non-white gestational surrogate (who gestates but does not contribute DNA to a child) than an egg or sperm donor who is not white.[79]

Prospective parents in search of donated gametes often describe their racial specifications in nonracial terms as a desire to produce children who look like them and who fit in with their broader extended families. Historically, the fertility industry emerged to help white couples expand their families. Clinics therefore tend to construct and market whiteness as "neutral" in the sense of being "unmarked, unencumbered by geographic and ethnic specificity."[80] Clients and brokers rarely need to state an explicit desire to procure gametes that will create children who can "pass" as the biological kin of white heterosexual parents because the notion is so naturalized as to go without saying.[81] Yet the illusion of whiteness as neutrality breaks down when fertility industry clientele expands beyond white couples and individuals. In some cases, prospective parents themselves want donors who share their phenotypic features and/or their racial or ethnic identity, but in other cases they do not.[82] Sometimes these desires come into conflict with one another—for example, when the donor who looks most like a prospective parent does not share their racial or ethnic identity.[83]

The commoditization of race in the fertility industry reveals that what looks like a set of consumer choices is in fact a market formed by existing preferences and prejudices and iteratively reinforced by gamete recipients and brokers.[84] Individuals and couples whose preferences are not aligned with this mainstream market have fewer choices available to them. Practitioners of fertility medicine have long structured the market for assisted reproductive technology in ways that exclude minoritized people. The legal scholar Dorothy Roberts observed over 25 years ago in her essential work *Killing the Black Body* that the fertility industry codes infertility as a white woman's disease in spite of much higher rates of infertility among African American women.[85] This coding both reflects and perpetuates the fact that, even though African American women are more likely to suffer from infertility, they are less likely to be able to access or afford assisted reproductive technologies. However, simply characterizing this inequality as a problem of access or money erases the very real presence of racism in fertility medicine encounters. The medical encounter itself enacts what Davis calls "obstetric racism," in which Black women are subject to racial and gender hierarchies that structure clinical relationships.[86] Finding Black gamete donors is difficult for prospective parents who prefer to do so. For example, of the 234 sperm donors available in the California Cryobank in July 2022, only three were listed as "Black or African American."

According to the legal scholar Camille Gear Rich, our observation is typical: many gamete agencies have no Black donors at all, and those that do have very few.[87] As a result, minoritized individuals wanting to donate gametes and couples seeking gametes from minoritized donors turn disproportionately to informal and unregulated networks of exchange.[88]

The desire for intrafamily racial sameness or blending to match mixed-race couples often rests on unspoken ideas of race as kinship. It is not universal, however. Deomampo has described both white and Asian prospective parents seeking out donors with the other racial identity (white parents seeking Asian donors and Asian parents seeking white donors) in order to create a child who is *hapa* (half-Asian and half-white) because such an identity is valued in Hawai'i.[89] Medical anthropologists working in Asia have also identified a desire for white sperm and egg donors. Prospective parents may describe this preference as an expression of their cosmopolitanism,[90] though the sociologist Amrita Pande contends that this purchasing of white privilege reflects and perpetuates the global valuation of whiteness above all other racial identities.[91]

Prevailing notions of racial identities as primarily phenotypic and genetic mean that they are sometimes construed as socially representative, which means they can be used in an instrumental way. For example, some white Evangelical Christians have championed transracial adoption of non-white children as a means of achieving racial reconciliation.[92] The anthropologist Risa Cromer reveals a recent trend among white Evangelicals to "adopt" non-white embryos (created during IVF but not used by the couple who created them) as a means of addressing "racial conflict."[93] Embedded in this practice is the notion that, somehow, biogenetic phenotype works as a stand-in for the cultural experience of race to such an extent that deep social rifts can be healed through transracial adoption. Evangelical Christians tend to see racial identity as biological and phenotypic. Even as they condemn racism, they characterize it as being only about skin color.

The chapter by the scholar of religion Meaghan O'Keefe, "Evangelical Christianity, Race, and Reproduction," explores white Evangelical Christian ideas of race and how it is reproduced. O'Keefe traces the historical relationship of race science to religion in the United States in order to contextualize contemporary religious and political beliefs about race. The slippage between social and genetic identities we discussed above figures slightly differently in white Evangelical communities. White Evangelicals tend to avoid discussions of racial identity. When such topics do come up, they are often viewed as divisive, and divisiveness is often cast in religious terms: to point out racial discrimination within the Church is to undermine Christ's vision for the Church as the unity of all believers.[94] More generally, Evangelical attitudes toward behavior genetics function in similar ways to Evangelical beliefs about racial politics. White Evangelicals are far more likely to attribute economic disparities between white and Black individuals to the result of poor personal choices. They see claims about racial discrimination as an excuse for not taking individual responsibility. The theme of individual responsibility carries into genetics, with white Evangelical leaders typically condemning using "your genes" as an excuse for bad behavior (including being gay). The one area in which biology and social roles are considered inseparable is gender. For white Evangelicals, men and women are complementary (they complete one another), and chromosomal sex is determinative.

As we have discussed throughout this introduction, race is widely understood to be genetically transmitted, even though it is maintained and experienced socially and institutionally. While some prospective parents may carefully curate the race of their offspring-to-be, they can't necessarily control how their children will identify or be identified once they are born. In the chapter "How Does a Baby Have a Race?," the public health geneticist Alice B. Popejoy examines how popular understandings of race and its transmission intersect with bureaucratic structures to assign racial identities to newborns. Social scientists and the US government typically define race as a category of identification that is co-constructed between an individual and the society in which they live. Popejoy describes how the process begins at or even before birth, when families, medical personnel, researchers, and governmental agencies apply racialized classifications to infants and their parents. Before the newborn has had an opportunity to develop a racialized sense of self, they are born into a context in which their parents' experiences and even their own prenatal ones are shaped by race, and race is assigned to them through bureaucratic and statistical processes in which they have no input.

CONCLUSION: CROSSING DISCIPLINARY BOUNDARIES

Ideas about how happy, healthy families ought to be formed are mediated by a fairly homogeneous set of institutional and commercial entities, and these entities shape what choices are made in the context of genetic disease and what ethnicity, race, and family resemblance mean. In the realm of reproductive and fertility medicine, people choose from a preordained set of options. These options are structured by racial categories that have been produced and maintained legally, socially, and scientifically over generations. New ideas about what constitutes populations and what ancestry is and is not have more recently developed in the context of genomic research and DNA testing. Popular, legal, historical, and scientific ideas about genetic and racial identities have commingled and combined, creating an amalgamation of sometimes conflicting ideas about who we are and how our self-identities and those identities forced upon us shape our experiences. These topics cross disciplinary boundaries; working in and with human genomics and genetics means thinking seriously about the social consequences of classifying race and ancestry and about distinguishing between health and disease and between favorable and unfavorable social outcomes. Similarly, humanistic scholarship on such topics needs grounding in scientific approaches, the workings of assisted reproductive technologies, and the intricacies of institutional biomedical research. None of this work can be done in isolation. This book offers a set of reflections and arguments that have developed from our conversations with one another. We hope to create similar opportunities for all our readers to think clearly and talk to one another about the fundamental questions we face together in the genomic age.

NOTES

1. "State-Specific Assisted Reproductive Technology Surveillance," Centers for Disease Control and Prevention, accessed February 8, 2024, https://archive.cdc.gov/#/details?url=https://www.cdc.gov/art/state-specific-surveillance/index.html.

2. Jonathan Boyarin, *The Unconverted Self: Jews, Indians, and the Identity of Christian Europe* (Chicago: University of Chicago Press, 2009).

3. J. Kameron Carter, *Race: A Theological Account* (New York: Oxford University Press, 2008).

4. Jennifer L. Hochschild and Brenna Marea Powell, "Racial Reorganization and the United States Census 1850–1930: Mulattoes, Half-Breeds, Mixed Parentage, Hindoos, and the Mexican Race," *Studies in American Political Development* 22, no. 1 (2008): 59–96; Melissa Nobles, *Shades of Citizenship: Race and the Census in Modern Politics* (Stanford, CA: Stanford University Press, 2000); Clara E. Rodriguez, *Changing Race: Latinos, the Census, and the History of Ethnicity in the United States* (New York: NYU Press, 2000).

5. Kenneth Prewitt, *What Is "Your" Race? The Census and Our Flawed Efforts to Classify Americans* (Princeton, NJ: Princeton University Press, 2013).

6. Brian Gratton and Emily Klancher Merchant, "*La Raza*: Mexicans in the United States Census," *Journal of Policy History* 28, no. 4 (2016): 537–67.

7. UNESCO, *Four Statements on the Race Question* (Paris: United Nations Educational, Scientific and Cultural Organization, 1969), 30, https://unesdoc.unesco.org/ark:/48223/pf0000122962/PDF/122962engo.pdf.multi.

8. Richard C. Lewontin, "The Apportionment of Human Diversity," in *Evolutionary Biology*, ed. Theodosius Dobzhansky, M. K. Hecht, and W. C. Steere (New York: Springer, 1972), 381–98.

9. "Office of Management and Budget Directive No. 15: Race and Ethnic Standards for Federal Statistics and Administrative Reporting," Centers for Disease Control and Prevention, accessed February 8, 2024, https://wonder.cdc.gov/wonder/help/populations/bridged-race/directive15.html.

10. Barbara A. Koenig, Sandra Soo-Jin Lee, and Sarah S. Richardson, eds., *Revisiting Race in a Genomic Age* (New Brunswick, NJ: Rutgers University Press, 2008), 3.

11. Agustín Fuentes, *Race, Monogamy, and Other Lies They Told You: Busting Myths about Human Nature*, 2nd ed. (Oakland: University of California Press, 2022).

12. Joan H. Fujimura, Deborah A. Bolnick, Ramya Rajaopalan, Jay S. Kaufman, Richard C. Lewontin, Troy Duster, Pilar Ossorio, and Jonathan Marks, "Clines without Classes: How to Make Sense of Human Variation," *Sociological Theory* 32, no. 3 (2014): 208–27.

13. Jonathan Marks, "Race: Past, Present, and Future," in *Revisiting Race in a Genomic Age*, ed. Barbara A. Koenig, Sandra Soo-Jin Lee, and Sarah S. Richardson (New Brunswick, NJ: Rutgers University Press, 2008), 21–38, 29.

14. Jessica P. Cerdeña, Vanessa Grubbs, and Amy L. Non, "Genomic Supremacy: The Harm of Conflating Genetic Ancestry and Race," *Human Genomics* 16 (2022): 18; Anna C. F. Lewis, Santiago J. Molina, Paul S. Appelbaum, Bege Dauda, Anna Di Rienzo, Agustín Fuentes, Stephanie M. Fullerton,

Nanibaa' A. Garrison, Nayanika Ghosh, et al., "Getting Genetic Ancestry Right for Science and Society," *Science* 376, no. 6590 (2022): 250–52.

15. Iain Mathieson and Aylwyn Scally, "What Is Ancestry?," *PLoS Genetics* 16, no. 3 (2020): e1008624.

16. Catherine Nash, *Genetic Geographies: The Trouble with Ancestry* (Minneapolis: University of Minnesota Press, 2015).

17. Graham Coop, "Genetic Similarity versus Genetic Ancestry Groups as Sample Descriptors in Human Genetics," arXiv (2022), https://arxiv.org/pdf/2207.11595.pdf.

18. Deborah A. Bolnick, Duana Fulwiley, Troy Duster, Richard S. Cooper, Joan H. Fujimura, Jonathan Kahn, Jay S. Kaufman, Jonathan Marks, Ann Morning, Alondra Nelson, Pilar Ossorio, Jenny Reardon, Susan M. Reverby, and Kimberly TallBear, "The Science and Business of Genetic Ancestry Testing," *Science* 318 (2007): 399–400.

19. Aaron Panofksy and Joan Donovan, "Genetic Ancestry Testing among White Nationalists: From Identity Repair to Citizen Science," *Social Studies of Science* 49, no. 5 (2019): 653–81.

20. Alondra Nelson, *The Social Life of DNA: Race, Reparations, and Reconciliation after the Genome* (Boston: Beacon Press, 2016); Panofsky and Donovan, "Genetic Ancestry Testing."

21. Joseph L. Graves Jr. and Alan H. Goodman, *Racism, Not Race: Answers to Frequently Asked Questions* (New York: Columbia University Press, 2021).

22. Terence Keel, *Divine Variations: How Christian Thought Became Racial Science* (Stanford, CA: Stanford University Press, 2018), 139.

23. Catherine Bliss, *Race Decoded: The Genomic Fight for Social Justice* (Stanford, CA: Stanford University Press, 2012); Nash, *Genetic Geographies*.

24. Fujimura et al., "Clines without Classes."

25. Coop, "Genetic Similarity versus Genetic Ancestry Groups."

26. Daniel Martinez HoSang, "On Racial Speculation and Racial Science: A Response to Shiao et al.," *Sociological Theory* 32, no. 3 (2014): 228–43; Jonathan Kahn, "'When Are You From?' Time, Space, and Capital in the Molecular Reinscription of Race," *British Journal of Sociology* 66, no. 1 (2015): 68–75; Lewis et al., "Getting Genetic Ancestry Right."

27. Kahn, "'When Are You From?'"

28. Rene Almeling, *Sex Cells: The Medical Market for Eggs and Sperm* (Berkeley: University of California Press, 2011).

29. Jennifer L. Eaton, Tracy Truong, Yi-Ju Li, and Alex Polotsky, "Prevalence of a Good Perinatal Outcome with Cryopreserved Compared with Fresh Donor Oocytes," *Obstetrics and Gynecology* 135, no. 3 (2020): 709–16.

30. Camille Gear Rich, "Contracting Our Way to Inequality: Race, Reproductive Freedom, and the Quest for the Perfect Child," *Minnesota Law Review* 104 (2019): 2375–469.

31. Debora Spar, *The Baby Business: How Markets Are Changing the Future of Birth* (Cambridge, MA: Harvard Business School Press, 2006).

32. David Plotz, *The Genius Factory: The Curious History of the Nobel Prize Sperm Bank* (New York: Random House, 2005).

33. Cynthia R. Daniels and Janet Golden, "Procreative Compounds: Popular Eugenics, Artificial Insemination and the Rise of the American Sperm Banking Industry," *Journal of Social History* 38, no. 1 (2004): 5–27.

34. Sarah Zhang, "The Fertility Doctor's Secret," *Atlantic*, April 2019, https://www.theatlantic.com/magazine/archive/2019/04/fertility-doctor-donald-cline-secret-children/583249/.

35. Daniels and Golden, "Procreative Compounds."

36. Daphna Birenbaum-Carmeli, Yoram Carmeli, and Sergei Gornostayev, "Researching Sensitive Fields: Some Lessons from a Study of Sperm Donors in Israel," *International Journal of Sociology and Social Policy* 28, no. 11/12 (2008): 425–39; Stanley Friedman, "Artificial Insemination with Donor Semen

Mixed with Semen of the Infertile Husband," *Fertility and Sterility* 33, no. 2 (1980): 125–28; Chia-Ling Wu, "Managing Multiple Masculinities in Donor Insemination: Doctors Configuring Infertile Men and Sperm Donors in Taiwan," *Sociology of Health and Illness* 33, no. 1 (2010): 96–113.

37. Laura Mamo, *Queering Reproduction: Achieving Pregnancy in the Age of Technoscience* (Durham, NC: Duke University Press, 2007).

38. Daniels and Golden, "Procreative Compounds."

39. Nellie Bowles, "The Sperm Kings Have a Problem: Too Much Demand," *New York Times*, January 8, 2021, https://www.nytimes.com/2021/01/08/business/sperm-donors-facebook-groups.html.

40. Paris Martineau, "Inside the Quietly Lucrative Business of Donating Human Eggs," *Wired*, April 23, 2019, https://www.wired.com/story/inside-lucrative-business-donating-human-eggs/.

41. Ilana Löwy, "How Genetics Came to the Unborn: 1960–2000," *Studies in History and Philosophy of Science Part C: Studies in History and Philosophy of Biological and Biomedical Sciences* 47 (2014): 154–62.

42. Shannon N. Conley, "Who Gets to Be Born? The Anticipatory Governance of Pre-Implantation Genetic Diagnosis Technology in the United Kingdom from 1978–2001," *Journal of Responsible Innovation* 7, no. 3 (2020): 507–27.

43. Alicia R. Martin, Masahiro Kanai, Yoichiro Kamatani, Yukinori Okada, Benjamin M. Neale, and Mark J. Daly, "Clinical Use of Current Polygenic Risk Scores May Exacerbate Health Disparities," *Nature Genetics* 51, no. 4 (2019): 584–91.

44. Alice B. Popejoy and Stephanie M. Fullerton, "Genomics Is Failing on Diversity," *Nature* 538 (2016): 161–64.

45. "GWAS Diversity Monitor," accessed February 8, 2024, https://gwasdiversitymonitor.com.

46. Mateus H. Gouveia, Amy R. Bentley, Thiago P. Leal, Eduardo Tarazona-Santos, Carlos D. Bustamante, Adebowale A. Adeyemo, Charles N. Rotimi, and Daniel Shriner, "Unappreciated Subcontinental Admixture in Europeans and European Americans and Implications for Genetic Epidemiology Studies," *Nature Communications* 14 (2023): 6802.

47. 1000 Genomes Project Consortium, "A Global Reference for Human Genetic Variation," *Nature* 526, no. 7571 (2015): 68–74.

48. Martin et al., "Clinical Use of Current Polygenic Risk Scores."

49. Sheetal Soni and Julian Savulescu, "Polygenic Embryo Screening: Ethical and Legal Considerations," Hastings Center, October 20, 2021, https://www.thehastingscenter.org/polygenic-embryo-screening-ethical-and-legal-considerations.

50. This shift from a focus on disease to a focus on risk maps onto the concept of biomedicalization described in Adele E. Clarke, Janet K. Shim, Laura Mamo, J. R. Fosket, and Jennifer R. Fishman, "Biomedicalization: Technoscientific Transformations of Health, Illness, and U.S. Biomedicine," *American Sociological Review* 68, no. 2 (2003): 161–94.

51. Simon Adler, "G: Unnatural Selection," *RadioLab*, July 25, 2019, https://www.wnycstudios.org/podcasts/radiolab/articles/g-unnatural-selection.

52. Cornelius A. Rietveld et al., "GWAS of 126,559 Individuals Identifies Genetic Variants Associated with Educational Attainment," *Science* 340, no. 6139 (2013): 1467–71; Aysu Okbay et al., "Genome-Wide Association Study Identifies 74 Loci Associated with Educational Attainment," *Nature* 533, no. 7604 (2016): 539–42; James J. Lee et al., "Gene Discovery and Polygenic Prediction from a Genome-Wide Association Study of Educational Attainment in 1.1 Million Individuals," *Nature Genetics* 50, no. 8 (2018): 1112–21; Aysu Okbay et al., "Polygenic Prediction of Educational Attainment within and between Families from Genome-Wide Association Analyses in 3 Million Individuals," *Nature Genetics* 54, no. 4 (2022): 437–49.

53. Patrick Turley, Michelle N. Meyer, Nancy Wang, David Cesarini, Evelynn Hammonds, Alicia R. Martin, Benjamin M. Neale, Heidi L. Rehm, Louise Wilkins-Haug, Daniel J. Benjamin, Steven Hyman, David Laibson, and Peter M. Visscher, "Problems with Using Polygenic Scores to Select Embryos," *New England Journal of Medicine* 385, no. 1 (2021): 78–86.

54. Sylvia H. Barcellos, Leandro Carvalho, and Patrick Turley, "The Effect of Education on the Relationship between Genetics, Early-Life Disadvantages, and Later-Life SES," National Bureau of Economic Research w28750 (2021), https://www.nber.org/papers/w28750; Callie Burt, "Challenging the Utility of Polygenic Scores for Social Science: Environmental Confounding, Downward Causation, and Unknown Biology," *Behavioral and Brain Sciences* 46 (2023): e207; Rosa Cheesman, Avina Hunan, Jonathan R. I. Coleman, Yasmin Ahmadzadeh, Robert Plomin, Tom A. McAdams, Thalia C. Eley, and Gerome Breen, "Comparison of Adopted and Non-Adopted Individuals Reveals Gene-Environment Interplay for Education in the U.K. Biobank," *Psychological Science* 31, no. 5 (2020): 582–91; Nicholas W. Papageorge and Kevin Thom, "Genes, Education, and Labor Market Outcomes: Evidence from the Health and Retirement Study," *National Bureau of Economic Research Working Paper* (2018), https://nber.org/papers/w25114.

55. For example, Kathryn Paige Harden, *The Genetic Lottery: Why DNA Matters for Social Equality* (Princeton, NJ: Princeton University Press, 2021); Robert Plomin, *Blueprint: How DNA Makes Us Who We Are* (Cambridge, MA: MIT Press, 2018).

56. Graham Coop and Molly Przeworski, "Lottery, Luck, or Legacy: A Review of *The Genetic Lottery: Why DNA Matters for Social Equality,*" *Evolution* 76, no. 4 (2022): 846–53.

57. "Premier Egg Donors," DonorNexus, accessed February 8, 2024, https://donornexus.com/services/egg-donation/premier-egg-donor-cycle.

58. For example, W. David Hill, Neil M. Davies, Stuart J. Ritchie, Nathan G. Skene, Julien Bryois, Steven Bell, Emanuele Di Angelantonio, David J. Roberts, Shen Xueyi, Gail Davies, David C. M. Liewald, David J. Porteous, Caroline Hayward, Adam S. Butterworth, Andrew M. McIntosh, Catharine R. Gale, and Ian J. Deary, "Genome-Wide Analysis Identifies Molecular Systems and 149 Genetic Loci Associated with Income," *Nature Communications* 10 (2019): 5741; Jorim J. Tielbeek, Ada Johansson, Tinca J. C. Polderman, Marja-Ritta Rautiainen, Philip Jansen, Michelle Taylor, Xiaoran Tong, Qing Lu, Alexandra S. Burt, Henning Tiemeier, et al., "Genome-Wide Association Studies of a Broad Spectrum of Antisocial Behavior," *JAMA Psychiatry* 74 (2017): 1242–50.

59. Andrea Ganna, Karin J. H. Verweij, Michel G. Nivard, Robert Maier, Robbee Wedow, Alexander S. Bush, Abdel Abdellaoui, Shengru Guo, J. Fah Sathirapongsasuti, 23andMe Research Team, et al., "Large-Scale GWAS Reveals Insights into the Genetic Architecture of Same-Sex Sexual Behavior," *Science* 365 (2019): 6456.

60. Dean H. Hamer, S. Hu, V. L. Magnuson, N. Hu, and A. M. Pattatucci, "A Linkage between DNA Markers on the X Chromosome and Male Sexual Orientation," *Science* 261, no. 5119 (1993): 321–27.

61. Nancy Ordover, *American Eugenics: Race, Queer Anatomy, and the Science of Nationalism* (Minneapolis: University of Minnesota Press, 2003).

62. Daniels and Golden, "Procreative Compounds"; Mamo, *Queering Reproduction.*

63. Ganna et al., "Large-Scale GWAS Reveals Insights."

64. Personal communication with Robbee Wedow, 2019. Prospective parents could, nevertheless, select embryos on the basis of their polygenic scores for same-sex sexuality.

65. For example, Joseph Vitti, "Opinion: Big Data Scientists Must Be Ethicists Too," *Broad Minded,* August 29, 2019, https://www.broadinstitute.org/blog/opinion-big-data-scientists-must-be-ethicists-too.

66. Amy Maxmen, "Controversial 'Gay Gene' App Provokes Fears of a Genetic Wild West," *Nature* 574 (2019): 609–10.

67. Turley et al., "Problems with Using Polygenic Scores."

68. Michelle N. Meyer, Tammy Tan, Daniel J. Benjamin, David Laibson, and Patrick Turley, "Public Views on Polygenic Screening of Embryos," *Science* 379, no. 6632 (2023): 541–43.

69. Robert Nozick, *Anarchy, State, and Utopia* (New York: Basic Books, 1974), 315n.

70. Dov Fox, "Racial Classification in Assisted Reproduction," *Yale Law Journal* 118, no. 8 (2009): 1844–99.

71. Amrita Pande, "Mix or Match? Transnational Fertility Industry and White Desirability," *Medical Anthropology* 40, no. 4 (2021): 335–47.

72. Daisy Deomampo, "Race, Nation, and the Production of Intimacy: Transnational Ova Donation in India," *Positions: East Asia Cultures Critique* 24, no. 1 (2016): 303–32.

73. Geoffrey C. Bowker and Susan Leigh Star, *Sorting Things Out: Classification and Its Consequences* (Cambridge, MA: MIT Press, 2000); Victoria Hattam, "Ethnicity and the Boundaries of Race: Rereading Directive 15," *Daedalus* 134, no. 1 (2005): 61–69.

74. Tessa Moll, "Making a Match: Curating Race in South African Gamete Donation," *Medical Anthropology* 38, no. 7 (2019): 588–602.

75. Seline Szupinski Quiroga, "Blood Is Thicker than Water: Policing Donor Insemination and the Reproduction of Whiteness," *Hypatia* 22, no. 2 (2007): 143–61.

76. Joe Mullin, "White Woman Sues Sperm Bank—Again—After Getting Black Man's Sperm," *Ars Technica*, April 25, 2016, https://arstechnica.com/tech-policy/2016/04/white-woman-sues-sperm -bankagainafter-getting-black-mans-sperm/.

77. Patricia J. Williams, "Babies, Bodies and Buyers," *Columbia Journal of Gender and Law* 33 (2016): 11–24, 16.

78. Rich, "Contracting Our Way to Inequality," 2397.

79. Kalindi Vora, "Indian Transnational Surrogacy and the Commodification of Vital Energy," *Subjectivity* 28, no. 1 (2009): 266–78.

80. Moll, "Making a Match," 593.

81. Pande, "Mix or Match?"

82. Alyssa Newman, "Mixing and Matching: Sperm Donor Selection for Interracial Lesbian Couples," *Medical Anthropology* 38, no. 8 (2019): 710–24.

83. Daisy Deomampo, "Racialized Commodities: Race and Value in Human Egg Donation," *Medical Anthropology* 38, no. 7 (2019): 620–33.

84. Michele Goodwin, "Reproducing Hierarchy in Commercial Intimacy," *Indiana Law Journal* 88 (2013): 1289–97; Lisa Chiyemi Ikemoto, "Reproductive Tourism," in *Beyond Bioethics: Toward a New Biopolitics*, ed. Osagie K. Obasogie and Marcy Darnovsky (Oakland: University of California Press, 2018), 339–49; Pande, "Mix or Match?"; Dorothy Roberts, *Fatal Invention: How Science, Politics, and Big Business Re-Create Race in the Twenty-First Century* (New York: The New Press, 2011); Camisha A. Russell, *The Assisted Reproduction of Race* (Bloomington: Indiana University Press, 2018); Natali Valdez and Daisy Deomampo, "Centering Race and Racism in Reproduction," *Medical Anthropology* 38, no. 7 (2019): 551–59; Williams, "Babies, Bodies and Buyers."

85. Dorothy Roberts, *Killing the Black Body: Race, Reproduction, and the Meaning of Liberty* (New York: Vintage Books, 1997).

86. Dána-Ain Davis, "Reproducing While Black: The Crisis of Black Maternal Health, Obstetric Racism and Assisted Reproductive Technology," *Reproductive Biomedicine & Society Online* 11 (2020): 56–64, 58.

87. Rich, "Contracting Our Way to Inequality."

88. Bowles, "Sperm Kings Have a Problem."

89. Deomampo, "Racialized Commodities."

90. Wei Wei, "Queering the Rise of China: Gay Parenthood, Transnational ARTs, and Dislocated Reproductive Rights," *Feminist Studies* 47, no. 2 (2021): 312–40.

91. Pande, "Mix or Match?"

92. Samuel L. Perry and Andrew Whitehead, "Christian Nationalism, Racial Separatism, and Family Formation: Attitudes toward Transracial Adoption as a Test Case," *Race and Social Problems* 7, no. 2 (2015): 123–34, 124.

93. Risa Cromer, "Making the Ethnic Embryo: Enacting Race in U.S. Embryo Adoption," *Medical Anthropology* 38, no. 7 (2019): 603–19.

94. Mark Driscoll, "Church: What Are the Characteristics of the Church?," Real Faith by Mark Driscoll, accessed February 8, 2024, https://realfaith.com/what-christians-believe/characteristics -church/.

DNA and Race

1

Are People like Metals?

Essences, Identity, and Certain Sciences of Human Nature

Mark Fedyk

PHILOSOPHICAL BACKGROUND

From Plato comes the seemingly eternal idea that people can be sorted and ranked as if they are metals: gold, silver, or brass and iron. This idea is introduced as an important political fiction in *The Republic*. Lest a city fall into disorder, its citizens must believe that all children are born with an inner metallic nature, which determines their public role or office. Children are to be told that their childhood was a dream; in reality, their nature was being formed deep in the earth by God, who then sent them up to the surface with false memories when they were ready to take their place in society as adults.

> Citizens, we shall say to them in our tale, you are brothers, yet God has framed you differently. Some of you have the power of command, and in the composition of these he has mingled gold [. . .]; others he has made of silver, to be auxiliaries; others again who are to be husbandmen and craftsmen he has composed of brass and iron; and the species will generally be preserved in the children. But as all are of the same original stock, a golden parent will sometimes have a silver son, or a silver parent a golden son. And God proclaims as a first principle to the rulers, and above all else, that there is nothing which they should so anxiously guard, or of which they are to be such good guardians, as of the purity of the race.[1]

It is a troubling phrase to read, that one, the words "purity of the race," even glossing over the issues of translation. Lisa Ikemoto's chapter will take up the concept of race purity in greater detail. But I want to stay with those words so as to use them anachronistically, and so use them to take us to a different place. Our next stop, specifically, is Locke and one of the enduring problems of empiricist philosophy of science. First, though, a bit more Plato.

> They should observe what elements mingle in their offspring; for if the son of a golden or silver parent has an admixture of brass and iron, then nature orders a transposition of ranks, and the eye of the ruler must not be pitiful towards the child because he has to descend in the scale and become a husbandman or artisan, just as there may be sons of artisans who having an admixture of gold or silver in them are raised to honour, and become guardians or auxiliaries. For an oracle says that when a man of brass or iron guards the State, it will be destroyed. Such is the tale; is there any possibility of making our citizens believe in it?[2]

The chapter by Carlos Andrés Barragán, Sivan Yair, and James Griesemer discusses the concept of admixture in its modern scientific guise. This chapter, however, is a critical examination of one way that science can be a source of credibility for origin stories about human nature. The scientific details of these modern stories are different, but the analogy is clear: the appeal of stories organized around the idea that people are like metals remains.

By "like metals," the idea here is not that people are to be valued in correspondence to the prices that precious metals have in markets for commodities. Rather, the idea is that both metals and humans have inner natures—"essences"—that determine their observable characteristics. Unlike the inner nature of the citizens of Plato's republic, however, inner natures—so we moderns have come to believe[3]—are not discoverable except by using specialized modes of inquiry. Only science now has the epistemological authority to tell stories about the inner natures—of metals, or of people, *if* people are like metals. If so, then stories that imply that human social categories like European or French or even Georgian may have genetic essences, analogous to how many people believe that metals like gold have atomic essences, may become common knowledge, just so long as the stories come from a place with sufficient scientific authority.

The distinction between *real essences* and *nominal essences* is central to Locke's philosophy of science, and it is useful here because it allows us a more refined set of distinctions than talking about inner natures.

The nominal essence of some category is an abstract mental representation that is shared by a group of people familiar with the perceptually characteristic properties of instances of that category. The contents of the nominal essence should all be observable properties—or, if not that, they should be properties that a person can more or less directly experience.

Real essences are not abstractions: they are the material, physical, or causal "stuff" out of which inductively useful (i.e., scientific) categories are composed. They are—depending on which flavor of metaphysics you want to endorse—the causal powers, the necessary and sufficient conditions, the essential properties, or the metaphysical grounds that make kinds or categories the kinds or categories that they are. Most importantly, real essences are hidden: they are not usually the things one can experience directly. Because of this, they must be discovered somehow, and for Locke, to a good first approximation, doing natural science is how we discover real essences.

A critical element of this picture is that nominal essences and real essences can be aligned or misaligned with one another. That is, we can form the hypothesis that a set of nominal essences N is "generated" by real essence R. But as a technical matter, the properties expressed in N cannot be the same properties expressed by R; otherwise, they would be the same category. But given that N and R express different properties, there is the problem of trying to discover some certainty-preserving technique or method for showing that R *really is* the grounds for—or foundation of, or cause of, or necessary for, or essence of—the properties of N. This is, as I mentioned, one of the enduring problems in empiricist philosophy of science.

About this problem a great deal has been written;[4] here, it suffices to say that Locke was mostly skeptical of the idea that a generally applicable technique or method could be found that solves the problem. Instead, the response Locke prefers goes like this:

> I would not here be thought to forget, much less to deny, that Nature in the Production of Things, makes several of them alike: there is nothing more obvious, especially in the Races of Animals, and all Things propagated by Seed. But yet, I think, we may say, the *sorting* of them under Names, *is the Workmanship of the Understanding, taking occasion from the similitude* it observes amongst them, to make abstract general *Ideas*, and set them up in the mind, with Names annexed to them, as Patterns, or Forms, . . . to which, as particular Things existing are found to agree, so they come to be of that Species, have that Denomination, or are put into that *Classis*.[5]

Nature makes things similar and different, but the kinds themselves are "the workmanship" of the mind. Real essences do not define natural kinds; natural kinds are social constructions. Natural kinds are human-made "conceptual tools" for thinking about "naturally produced" patterns observable to most people.

All the same, one of the stories contemporary scientists like to tell about science is that scientists routinely do achieve what Locke was skeptical of—specifically, discover the real essences that explain, cause, generate, or are otherwise responsible for certain nominal essences.[6] Which is to say: many scientists believe that they discover natural kinds by discovering real essences, the definitions of which then explain certain nominal essences—that, for example, gold "really just is" atoms with 79 protons in their nuclei. The real essence of gold is the *pure* essence of gold, one might say. Having a nucleus with 79 protons is the inner nature of gold.

Thus, we see in scientists' quest for molecular causes of various observable social patterns an updated version of the search for real essences. But the popularity and appeal of this story about how scientific discovery works does not address the epistemological ambiguity that is arguably the root of Locke's skepticism about whether real essences can ever define natural kinds. To put the argument rhetorically, why think that nominal essences are usually organized in some coherent metaphysical relation with real essences? If the relationship between real and nominal essences were straightforward, why would it take so much effort and energy to discover that gold "really is" anything that has the atomic number 79? But if the relationship between nominal and real essences were not straightforward, why should nominal essences be a guide to what real essences there are? Why care about nominal essences at all? Most observable gold is not elemental; indeed, most of the useful "nominal" kinds of gold are alloys, and so do not correspond at all to the "real" kinds given on the traditional periodic table.[7]

This ambiguity—whether we can ever know that some real essence is the "inner nature" of certain nominal essences—is what this chapter is about. Specifically, it provides a reading of the work of 23andMe and some relevant scientific prehistory that reinforces the thesis that, for all the technical sophistication of modern population genetics, ambiguity remains about the "origin stories" about humans that are suggested by the company's genetic analysis of ancestry and some of the social categories that people identify with. Indeed, allow me to introduce an explicit thesis: let us say that some schema or system of categories that expresses nominal essences has *Lockean ambiguity* if it is uncertain which, if any, real essences explain, cause, or otherwise ground the categories in the schema or the system. The intended conclusion of this chapter, then, is that the genetic analysis of human social categories offered by 23andMe cannot succeed in surmounting Lockean ambiguity about these categories.

The reason this argument matters is that maintaining Lockean ambiguity about human social categories is about as close to an ethical imperative as they come for us moderns. Reviewing evidence for this claim is beyond the scope of this chapter; Kwame Anthony Appiah's writings are a good place to start.[8] But if you share the unease about phrases like "purity of the race," then this evidence is probably already familiar to you. We should not presume that, for every nominal essence used to group, categorize, or act as a source of identity for people, there is a real essence to be found.

NEITHER METAL NOR ALLOY: RAZA ROUSTAM

Why? Humans are not metals. Not even alloys. Evidence for this is induction over human history: it is hard to impossible to find examples of the social categories that people identify with—whether by choice or by force or by parentage or by some other means—and that cannot be combined and recombined with one another without any limit over the course of an individual's life. This matters because it falsifies Plato's story: for Plato, people cannot change their inner metallic nature after they are born, and it is one's ancestry (not one's "nominal essences") that determines one's metallic nature, and thus one's station in society.

But, again, humans are not metals, and this fact can be illustrated more concretely by the story of Raza Roustam. Roustam is known to history through his association with Napoleon, a relationship that began soon after Napoleon landed in Egypt in 1798. Roustam remained connected with Napoleon until Napoleon's first loss of formal political power in France; these and other details of Roustam's life are collected in a memoir he wrote later in life.[9]

Roustam was born in Tiflis, in either 1781 or 1783, of Armenian parents. At the time, Tiflis was a part of the nominally independent Georgian kingdom of Kartli-Kakheti, though it was in 1783 that Tiflis fell under suzerainty of the Russian Empire, ending several centuries of de facto and de jure Persian rule.

As a young boy, Roustam escaped kidnapping several times by slavers before being successfully kidnapped and forced into slavery at age 13. His kidnapping followed centuries of tradition in the area, according to which young boys were taken from the Caucasus and sold into service as mamluks. The mamluks were originally raised as a fighting force in the seventh century; by the twelfth century, they formed an elite class of warriors and statesmen who held considerable political power throughout the Middle East. Though Armenian by birth, Roustam learned, as he was traded, that part of his value was contingent on his buyer's believing that he was Georgian. He wrote, "The Georgians and Mingrelians were preferred when it came to recruiting mamluks. I don't know why, because the Armenians are braver than any other people."[10]

Roustam consequently adopted a practice of passing as Georgian. He eventually arrived in Cairo, where he received his training as a mamluk, and where he then entered the service of Salih Bey, who was assassinated at about the time Napoleon's forces landed at Rosetta. Desiring to remain a mamluk, who by convention must have a master, rather than start life anew as a free person, Roustam sought out and soon thereafter was accepted into the service of a sheikh who had sworn loyalty to Napoleon. Roustam was then gifted to Napoleon by this sheikh, and Napoleon took Roustam to become his personal bodyguard and second valet.

Napoleon orientalized Roustam, calling him "Ali." Roustam was proud of being a mamluk; he frequently expressed pleasure and satisfaction in being able to dress in the ceremonial clothes of a mamluk. All the same, there are also few things more "French"—recognizing, of course, that it is hardly a static or univocal category—than a personal association with Napoleon. But after Napoleon was first deposed, despite having acquired a degree of fame in France, Roustam left Napoleon's service and lived out the rest of his life in France as a veteran of the Napoleonic wars. He died in 1845.

What is the relevance of this story? Most people's lives resemble Roustam's life. That is to say: none of us is born preconfigured to fit into the different social (that is, cultural, political, and moral, etc.) categories ("nominal essences") that are the source of life's opportunities and limitations. Whether just to survive, or to grow, or even to flourish, we all must find a way of adjusting, adapting, or conforming to the innumerable categories that give structure to the social worlds we move through. Frequently this means "taking on"—internalizing, or at least passing as a member of—categories (again, "nominal essences") that in no important sense we are born into, or have much prior practice living with. The relevance of Roustam's biography is therefore quite simple. He survived, grew, and eventually flourished by constructing a life that combined Armenian, Georgian, mamluk, oriental, Egyptian, enslaved, freedman, and French identity categories.

IDENTITY CATEGORIES

But what then are identity categories? As noted, they are, technically, nominal essences. But they are also the social categories that a person can inhabit, or at least conform to, through an exercise of their own agency, so that at least the appearance of being a member of a type or category of person becomes a practical possibility. Identity categories are different from the more familiar notion of social or cultural stereotypes.[11] Stereotypes are attributed to people in order to explain or understand or make predictions about them. Stereotyping—the action of attributing a category to a person, without regard to whether the person in question wants that category to be applied to them—can be a source of identity formation.[12]

But the focus here is not on how people conceptualize the identity of other people. Instead, the focus is on how a person *qua individual* relates, through their own agency, to the categories that give common structure to their inner mental life and outer social life. While it is, of course, the case that some, many, or even most of these categories may be foisted upon a person, even in such cases there is still the ongoing work of consciously adjusting one's psychobiography and psychosocial presentation to the reality of these categories—for instance, Roustam's insight that passing as Georgian was in his practical interest. Identity categories are those categories that a person has—at least partially, at least imperfectly—functionally reconciled with the rest of their psychobiographical and psychosocial self-understanding *and* the conventions, norms, mores, and habits of the social worlds they inhabit.[13] Identity categories are therefore ultimately by-products of widespread patterns of individual choice and agency, even if they sometimes have the appearance of being entirely structural or historical features of large groups of people.

Identity categories are also nominal essences par excellence. They are the workmanship of the understanding: we collectively imagine and construct and define and stipulate and feel these categories into existence, and to the extent that our thoughts, emotions, actions conform to the public dimensions of the categories, their existence becomes part of the fabric of human history. There are obvious and not-so-obvious social patterns associated with the categories. True, identity categories have a psychological basis,[14] but that is quite a different claim than asserting that certain real essences are the "naturally produced" hidden source of configuration or organization of any of our identity categories.

REAL ESSENCES: RONALD FISHER

It is characteristic of Enlightenment theories of human potential that they rest on certain strong assumptions about human nature—that there are certain "real essences" that either do in fact organize (or could, if things were different, be used to organize) identity categories.[15]

But following Darwin, and in particular his philosophy of emotions,[16] it becomes possible to use the logic of natural selection to try to discover real essences of human nature. With this shift, ancestry and descent are sometimes thought to determine the properties of a person's real essence, much as they do in Plato's myth. Ronald A. Fisher's program for eugenics is an example of this convergence; it is probably the most sophisticated modern version of the Platonic myth expressed using Darwinian logic.[17] Other aspects of eugenics will be discussed in greater detail in the chapters by Lisa Ikemoto, Emily Klancher Merchant, and Meaghan O'Keefe.

Consider, for example, how Fisher's 1919 article, "The Correlation between Relatives on the Supposition of Mendelian Inheritance," begins:

> Several attempts have already been made to interpret the well-established results of biometry in accordance with the Mendelian scheme of inheritance. It is here attempted to ascertain the biometrical properties of a population of a more general type than has hitherto been examined, inheritance in which follows this scheme. It is hoped that in this way it will be possible to make a more exact analysis of the causes of human variability. The great body of available statistics show us that the deviations of a human measurement from its mean follow very closely the Normal Law of Errors, and, therefore, that the variability may be uniformly measured by the standard deviation corresponding to the square root of the mean square error.[18]

Nature makes humans similar and different. But the real essences that are the causes of human biometrical variability can be discovered through the techniques of applied statistics.

The remainder of the article works out the mathematical foundations of what eventually became analysis of variance, or ANOVA. This technique does exactly what Fisher suggests: it allows you to calculate the constituent percentages of the total variance of some trait in some well-defined population that can be attributed to independent underlying causes of variance. To illustrate this technique, Fisher analyzes height, which is a nominal essence, and which of course can be expressed as a continuous variable. This appears to be one of the first examples in Fisher's work of what he calls "quantitative characteristics"—that is, those human traits that can be explained, at least in principle, by association with population-based measures of the frequencies of genetic values. To simplify, genes—human "inner nature"—explain human biometric variability, variability that is expressed in categories that are, technically, nominal essences.

But it is a significant leap to go from analyzing genetic patterns that explain biometrical variability in populations to treating genes as the real essence for human social categories. Nevertheless, this was a leap that Fisher believed would sooner or later be scientifically feasible. He was prepared to apply the concept of a quantitative characteristic to, seemingly, "all human problems":

> Our practical interest in the well-being of human populations turns predominantly on what are known as quantitative characters, such as exhibit themselves in intelligence tests, or in resistance to disease. What matters here, above everything, are the agencies which are capable of influencing the average of the population in a desirable or an undesirable sense. We are, therefore, much concerned with the theoretical and practical study of quantitative inheritance, with cases in which many Mendelian factors contribute to a single measurable effect, an aspect of genetic study which, owing to its difficulty, has been avoided in most centres of genetic research, but which plays such a central part in all human problems that, with us, it must constitute a major objective.[19]

Fisher appears to have hoped that enough of the human phenotype would comprise quantitative characteristics.[20]

From this hope, I want to suggest the following gloss on Fisher's eugenicist social philosophy. If human social categories can generally be associated with quantitative characteristics, then it may be possible to discover the real essences that shape, explain, cause, or otherwise ground such categories. These insights can then be used to better organize otherwise mysterious or messy or irrational aspects of various human social worlds.

FROM FISHER TO 23ANDME

Fisher did not seem to explicitly contemplate the idea that identity categories specifically could be treated as if they are quantitative characteristics. But this idea—again, that human identity categories can be treated as quantitative characteristics, and thus their real essence potentially limned by genetic analysis—appears to be central to the business model of 23andMe.

Below is an excerpt from the pitch letter that Anne Wojcicki, the CEO of 23andMe, sent to potential investors in 2007.

> Why do some people love to jump out of airplanes and some are terrified to fly? Why do some family members get diseases while others don't? The answers to these and other questions about human traits lie partially in our DNA. . . . 23andMe will enable consumers to have a better understanding of their ancestry and genealogy. Most people possess a natural curiosity of who they are, where they came from, and who their ancestors were. The answers to these and other questions about human traits lie partially in our DNA. The mission of 23andMe is to provide individuals access to their personal genetic data with the goal of unraveling some of these puzzles of inheritance.[21]

Nature makes humans similar and different. But the real essences that are the causes of human variability can be discovered through the analysis of personal genetic data.

So, the leap here is the same as it was for Fisher: there is the hope that enough of the subjectively interesting aspects of human variability can be analyzed as quantitative characteristics. Consider thus 23andMe's effort "to further our understanding of the genetics of musicality." Musicality is treated as a composite construct formed by weighting a set of quantitative measures: "self-reported beat synchronization ability . . . and objectively measured rhythm discrimination" as well as starting age of playing music, amount of musical practice, a psychometric measure of flow proneness.[22] Rhythm discrimination, for instance, appears to be mediated by assortative mating in certain Scandinavian populations.

But what is perhaps most innovative about 23andMe's social philosophy is the construction of a set of novel identity categories that are, by design, quantitative characteristics. Rather than trying to discover a set of historically independent identity categories that are also quantitative characteristics, 23andMe has developed its own inventory. These categories resemble identity categories that are ethnographic common knowledge in many Western societies; technically, they refer only to reference populations for the purpose of calibrating models that predict ancestry from samples of DNA. But they mostly take the names of either contemporary political groupings or commonly known ethnic groups. To determine someone's ancestry, a sample of that individual's DNA is projected into these social categories using an SVM algorithm. The social categories are nested, as depicted in figure 1.1.[23]

A person's ancestry is some combination of the outermost cells, adding up to 1 or 100 percent, so someone could be 47 percent "Arabia," 41 percent "Melanesia," and 12 percent "Kerala." The implied invitation here is straightforward: since the genetic information is categorized using the 23andMe social categories, so, too, presumably, is the person who supplied the genetic

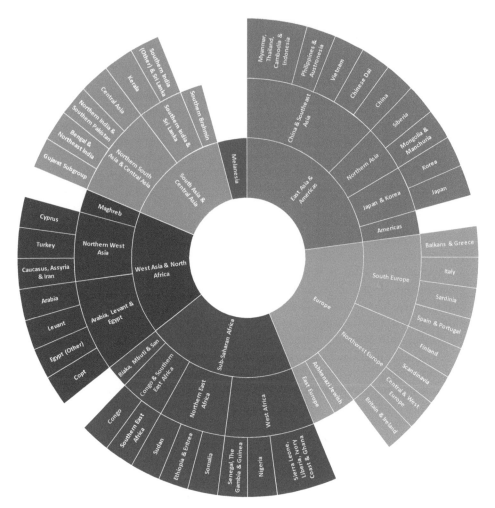

FIGURE 1.1. Early (ca. 2014) 23andMe reference categories. Image created by the author.

information—if, that is, they begin to treat the 23andMe social categories as identity categories.

IDENTITY CATEGORIES, QUANTITATIVE CHARACTERISTICS, REAL ESSENCES, AND NOMINAL ESSENCES

But why would anyone want to do this? I suggest that one plausible explanation is the belief that science discovers real essences that explain nominal essences.[24] "Being scientific" is central to 23andMe's public identity, and so one gloss on 23andMe's occasional marketing slogan—that they offer clients a way to "know your personal story, in a whole new way"—is that 23andMe can provide you with the real essences (by categorizing a sample of your genes) that explain the nominal essences that you may identify with—specifically, any of your own preexisting identity categories that coincide with at least some members of a set of 23andMe social categories that is projected from 23andMe's categorization of your genes.

Note that this works so long as users of 23andMe's services are prepared to make a similar leap that Fisher makes in expanding his concept of quantitative characteristics. Almost anything can be measured using a quantitative scale or instrument, and there is no reason why some large groups of people could not identify with social categories that are, technically, quantitative characteristics. But in fact, most people do not do this. Most—probably all—of the social categories that become identity categories are nominal categories in the sense of levels of measure.[25] They are not, that is to say, technically, quantitative characteristics— that is, variables that take either integers or real numbers as values and that can therefore be subject to mathematical operations.

The argument for this takes us back to Roustam. The elements of Roustam's social identity include the nominal categories ("nominal essences") French, Georgian, Armenian, and mamluk. None of these are quantitative characteristics: it makes no sense to express these social categories using scales built from rational or real numbers. Roustam was not 47 percent French and 41 percent Georgian and 12 percent Armenian. Instead, as his autobiography celebrates, these categories are nonexclusively aggregative over the course of his life's history. Becoming French made Roustam no less and no more mamluk and no less and no more Georgian.

So there is a gap between 23andMe's social categories and the identity categories for most people. Genetic categories might be the real essences for the former, but they are not automatically real essences for the latter.

Indeed, we can briefly examine the three metaphysical options for linking between 23andMe's genetic categories and people's (usually preexisting) identity categories as a way of strengthening this observation, for what this examination shows is that the metaphysics of the relevant categories will not close this gap. Thus, let R be the set of "real essences" that is given by 23andMe's categorization of a sample of genetic information, and let N be the set of "nominal essences" that expresses the set of identity categories for the same person from which the genetic sample was drawn. (N is therefore not the social categories that 23andMe projects a sample of genetic information into.)

The strongest relationship between N and R is that of identity, such that N reduces to R because $N = R$.[26] We can ignore this because the number of categories in the R for 23andMe is vastly fewer than the number of categories in any person's N. The residual Ns would be left unexplained. But as a technical matter, a nominal category cannot be mathematically or logically identical to an interval or a ratio measure.

The same observation rules out a slightly weaker metaphysical connection— namely, the assumption that N and R have the same formal structure.[27] Technically, real-world identity categories are aggregative without being additive: when an Armenian moves to France, they do not thereby become proportionally less Armenian and more French; they are, sooner or later, both Armenian and French. But the 23andMe social categories do not behave this way: if we somehow splice new genes into someone's genome that map into certain 23andMe social categories, this would cause a proportional decrease in the percentages of the other 23andMe social categories.

The only remaining metaphysical assumption about the relationship between N and R is that R is the cause of the various Ns.[28] But these Rs are either just too far in the past or too few in kind to be the causes of most of the relevant Ns—that is, the many different social categories that people come to identify with.

So, why would someone think that 23andMe's social categories are relevant to their identity, as Anne Wojcicki appears to hope? It seems, perhaps ironically, that, if it is part of one's identity "to be scientific" and this is taken to mean that it is important to try to discover the real essences that explain the nominal essences that are one's identity categories, then 23andMe has something to offer. They can provide an origin story a bit like Plato's myth for anyone with such a scientistic orientation: they provide a set of (novel) social categories that some people can choose to identify with.

CONCLUSION: PEOPLE ARE NOT LIKE METALS

At this point, we can leave science and return to ethics. People are not like metals—just so long as they do not adopt epistemological values that lead them to internalize as identity categories *only* categories that are, technically, quantitative characteristics that can be defined or explained genetically, and where the explanation comes from a source with sufficient scientific authority.

This is why it matters that we see the principle of Lockean ambiguity as an important moral imperative. Construed this way, it functions as a guardrail against trying to discover the real essences that somehow account for or explain human social categories. This is not the same as saying that these categories cannot be explained scientifically, of course. History, anthropology, sociology, folklore, and religious practice are all sources of science or science-like knowledge about these categories. The technical point is that treating Lockean ambiguity about social categories as a moral imperative prevents us from trying to explain away social diversity by reducing it to something else.

People are not metals; thus, it does not make sense to ask what they purely are, when this is a question about what a person's "real essence" is, asked because of some kind of concern about what social categories a person can be or should be included within. It is a moral error to ask whether Roustam was really Armenian or Georgian or mamluk or French—a moral mistake, that is, to search for some real essence that can explain what nominal categories "really" were his to identify with.

NOTES

1. Plato, *The Republic* (New York: Modern Library, 1960).

2. Plato, *Republic*.

3. Bruno Latour, *On the Modern Cult of the Factish Gods* (Durham, NC: Duke University Press, 2010).

4. R. Boyd, "Kinds as the 'Workmanship of Men': Realism, Constructivism, and Natural Kinds," in *Rationalität, Realismus, Revision*, ed. J. Nida-Rümelin (Internationalen Kongresses der Gesellschaft für Analytische Philosophie, 1999), 52–89; R. Boyd, "Rethinking Natural Kinds, Reference and Truth: Towards More Correspondence with Reality, Not Less," *Synthese* 198 (2021): 2863–903; Y. Onishi and D. Serpico, "Homeostatic Property Cluster Theory without Homeostatic Mechanisms: Two Recent Attempts and Their Costs," *Journal for General Philosophy of Science* 53 (2022): 61–82; T. Lombrozo, "Causal-Explanatory Pluralism: How Intentions, Functions, and Mechanisms Influence Causal Ascriptions," *Cognitive Psychology* 61 (2010): 303–32; C. F. Craver, "Mechanisms and Natural Kinds," *Philosophical Psychology* 22 (2009): 575–94; Hilary Putnam, *The Threefold Cord: Mind, Body, and World* (New York: Columbia University Press, 1999).

5. John Locke, *An Essay concerning Human Understanding* (Indianapolis: Hackett Publishing, 1996).

6. Lorraine Daston and Peter Galison, *Objectivity* (Princeton, NJ: Princeton University Press, 2007).

7. Michael D. Gordin, *A Well-Ordered Thing: Dmitrii Mendeleev and the Shadow of the Periodic Table*, rev. ed. (Princeton, NJ: Princeton University Press, 2018).

8. Kwame Anthony Appiah, "Reconstructing Racial Identities," *Research in African Literatures* 27 (1996): 68–72.

9. Raza Roustam, *Napoleon's Mameluke: The Memoirs of Roustam Raeza* (New York: Enigma Books, 2015).

10. Roustam, *Napoleon's Mameluke*.

11. M. Rhodes, "How Two Intuitive Theories Shape the Development of Social Categorization," *Child Development Perspectives* 7 (2013): 12–16; E. Foster-Hanson and M. Rhodes, "Stereotypes as Prototypes in Children's Gender Concepts," *Developmental Science* 26 (2023): e13345.

12. Ian Hacking, "Making Up People: Clinical Classifications," *London Review of Books* (2006), https://www.lrb.co.uk/the-paper/v28/n16/ian-hacking/making-up-people.

13. P. Y. Gal'perin, "On the Notion of Internalization," *Soviet Psychology* 5 (1967): 28–33; P. M. Bromberg, "Shadow and Substance: A Relational Perspective on Clinical Process," *Psychoanalytic Psychology* 10 (1993): 147–68.

14. Mark Fedyk, *The Social Turn in Moral Psychology* (Cambridge, MA: MIT Press, 2017).

15. For example, Lynda Lange reads Rousseau as a feminist critiquing early capitalist society for establishing identity categories that distort the expression of amour de soi, itself a universal component of, and thus an element of the real essence of, human nature. Lynda Lange, "Rousseau and Modern Feminism," *Social Theory and Practice* 7 (1981): 245–77.

16. Fedyk, *Social Turn in Moral Psychology*.

17. For an interesting comparison with Fisher, see the somatotype theory of Earnest A. Hooton. N. Rafter, "Earnest A. Hooton and the Biological Tradition in American Criminology," *Criminology* 42 (2004): 735–72; J. E. L. Carter and B. H. Heath, "Somatotype Methodology and Kinesiology Research," *Kinesiology Review* 10 (1971): 10–19; E. A. Hunt and W. H. Barton, "The Inconstancy of Physique in Adolescent Boys and Other Limitations of Somatotyping," *American Journal of Physical Anthropology* 17 (1959): 27–35.

18. Ronald A. Fisher, "XV.—The Correlation between Relatives on the Supposition of Mendelian Inheritance," *Earth and Environmental Science Transactions of the Royal Society of Edinburgh* 52 (1919): 399–433.

19. Ronald A. Fisher, "Eugenics, Academic and Practical," *Eugenics Review* 27 (1935): 95–100.

20. It is important not to conflate the measure (or a construct that expresses information generated by a measure) with the mechanism responsible for inducing the relevant variability. To introduce a point I will return to below, almost anything can be measured quantitatively; the real question is whether the constructs yielded by any such measures can be used to make sufficiently accurate predictions about trends in various populations. P. E. Meehl, "Theory-Testing in Psychology and Physics: A Methodological Paradox," *Philosophy of Science* 34 (1967): 103–15; P. E. Meehl, "Theoretical Risks and Tabular Asterisks: Sir Karl, Sir Ronald, and the Slow Progress of Soft Psychology," in *The Restoration of Dialogue: Readings in the Philosophy of Clinical Psychology*, ed. R. B. Miller (Washington, DC: American Psychological Association, 1992), 523–55. Success in the first endeavor is no guarantee of success in the second. D. A. Freedman, *Statistical Models and Causal Inference: A Dialogue with the Social Sciences* (New York: Cambridge University Press, 2010).

21. Anne Wojcicki, "Letter to Investors," 2007, accessed March 18, 2024, https://assets.bwbx.io/images/users/iqjWHBFdfxIU/i_JWmL6tfMhA/v2/1650x2200.jpg.

22. L. W. Wesseldijk et al., "Using a Polygenic Score in a Family Design to Understand Genetic Influences on Musicality," *Scientific Reports* 12 (2022): 14658; F. Dudbridge, "Power and Predictive Accuracy of Polygenic Risk Scores," *PLoS Genetics* 9 (2013): e1003348.

23. Adapted from Eric Y. Durand, Chuong B. Do, Peter R. Wilton, Joanna L. Mountain, Adam Auton, G. David Poznik, and J. Michael Macpherson, "A Scalable Pipeline for Local Ancestry Inference Using Tens of Thousands of Reference Haplotypes," bioRxiv (2021), https://doi.org/10.1101/2021.01.19.427308.

24. J. M. Vienne, "Locke on Real Essence and Internal Constitution," *Proceedings of the Aristotelian Society* 93 (1993): 139–53; E. J. Lowe, "Locke on Real Essence and Water as a Natural Kind: A Qualified Defense," *Aristotelian Society Supplementary Volume* 85 (2011): 1–19; S. Goodin, "Why Knowledge of the Internal Constitution Is Not the Same as Knowledge of the Real Essence and Why This Matters," *Southwest Philosophy Review* 14 (1998): 149–55.

25. P. Velleman and L. Wilkinson, "Nominal, Ordinal, Interval, and Ratio Typologies Are Misleading," in *Trends and Perspectives in Empirical Social Research*, ed. Ingwer Borg and Peter P. Mohler (Berlin: De Gruyter, 2011), 161–77; Cristian Larroulet Philippi, "On Measurement Scales: Neither Ordinal Nor Interval?," *Philosophy of Science* 88 (2021): 929–39.

26. Paul Oppenheim and Hilary Putnam, "Unity of Science as a Working Hypothesis," in *Concepts, Theories, and the Mind-Body Problem: Minnesota Studies in the Philosophy of Science*, ed. Herbert Feigl, Michael Scriven, and Grover Maxwell (Minneapolis: University of Minnesota Press, 1958), 3–36; R. Kirk, "Nonreductive Physicalism and Strict Implication," *Australasian Journal of Philosophy* 79 (2001): 544–52.

27. J. Ladyman, "What Is Structural Realism?," *Studies in History and Philosophy of Science* 29 (1998): 409–24; O. Bueno, "Structural Realism, Mathematics, and Ontology," *Studies in History and Philosophy of Science* 74 (2019): 4–9.

28. L. N. Ross, "Multiple Realizability from a Causal Perspective," *Philosophy of Science* 87 (2020): 640–62; C. Gillett, "The Metaphysics of Realization, Multiple Realizability, and the Special Sciences," *Journal of Philosophy* 100 (2003): 591–603; D. S. Brooks, J. DiFrisco, and W. C. Wimsatt, *Levels of Organization in the Biological Sciences* (Cambridge, MA: MIT Press, 2021).

2

A Colorful Explanation

Promoting Genomic Research Diversity Is Compatible with Racial Social Constructionism

Tina Rulli

This chapter explores the possible tension between the call for more diversity in genomic research and the view that races are socially constructed and not biologically real. Does the claim that we need more diversity in genomic research, often understood in racial terms, rely upon an explicit commitment to biological race realism?

Proponents of genomic medicine hope to employ associations between gene variants and disease states and drug metabolism to predict, diagnose, or treat disease in individuals through genetic testing, including in preimplantation genetic diagnosis and prenatal screenings, and to develop targeted gene therapies or interventions. Genomic medicine relies upon genome-wide association studies (GWAS), where individual genomic samples are assessed and compared for patterned associations between known gene variants and disease states or drug responses. The targets of GWAS are usually complex diseases, those associated with multiple genes. Since the effects of each gene may be tiny, GWAS requires databases of genomic samples from a very large number of individuals to sufficiently power the associations. Currently, however, individuals of primarily European descent are vastly overrepresented in GWAS. The GWAS Diversity Monitor, which tracks real-time diversity statistics for participants, reports that 95.05 percent of participants are of European descent, with slightly more than 3 percent of Asian descent.[1]

There is a widespread call to racially and ethnically diversify genomic research.[2] Proponents of diversification claim that population diversity—often described at the continental level, echoing familiar continental conceptions of race—is needed to ensure the accuracy of genomic medicine and to extend the benefits of genomic medicine to all people.

The call for racial diversity in genomic research might imply that race must be biologically real, that race is encoded at the genetic level. Why else would racial inclusion be important in genomic research? The call for diversity may also seem to imply that differently racialized people have different genes. But neither claim is true. Here, I argue that racial diversity efforts in genomic research are compatible with the denial of a biological reality for race and compatible with social constructionism about race. Thus, genomic researchers advocating for racial diversity in genomic research need not be committed to or seen as advocating for the view that racial categories are biologically real.

A few disclaimers at the outset. I am not advocating for genomic medicine. The majority of race-based differences in disease have socioenvironmental explanations.[3] Nor do I think increasing racial diversity is the *best* way to go about increasing genomic diversity. The use of genetic similarity, a continuous measure based on genes themselves, would better ensure representation of human population diversity. But if we take geneticists at their word—that genomic medicine will bear fruit—it is incumbent upon us that these putative benefits be equitably distributed. The calls for inclusivity in genomic research often take the form of racial diversity. I argue that it is not incoherent to advocate for racial diversity in genomic research and to embrace the dominant, most defensible view of what racial categories are, the social constructionist view. That is, one is mistaken if one sees these calls for racial diversity as requiring the truth of biological race realism. Instead, a call for racially diversifying genomic research can be a practical strategy in the just allocation of benefits across diverse people, even among social constructionists about race.

In what follows, I will center the US conception of race, which identifies five races pertaining to five continents. In this conception, the categories are white (European descent), Black (African descent), Asian, Pacific Islander, and Indigenous American.[4] This continental race-based classification is widely adopted by geneticists and invoked even when not talking directly about race—for example, when making population or ancestral group assignments for people.

I face a difficulty in citing studies that use population descriptors, referring to *race, descent, genetic ancestry*, or continental level *populations*. There is a lack of consistency among scientists in the use of these terms, and, further, these groupings are typically given at the continental level, reifying the idea that there are meaningful biological groupings that map onto our conventional notion of race. But it is this very idea that I am arguing against here. Recent, prominent efforts have been made to scrutinize descent-based descriptors and to render their usage more consistent, intentional, and transparent. In 2023 the National Academies of Sciences, Engineering, and Medicine (NASEM) issued a report whose mission is to clarify the use of group labels for individual research participants in scientific studies out of concern with the unstandardized, unscientific use of racial or ethnic categories in population descriptors.[5] The committee does not recommend terms of use; rather, it outlines a shared approach to the use of population descriptors in accordance with the principles of respect, beneficence, equity, justice, and transparency, among other values. This chapter is part of a critical literature on these race-based concepts. In citing or referring to studies, I am not advocating for the use of these race-based terms. But were I to change the nomenclature these studies employ, I might change their intended meaning, whatever it is. Thus, I have opted to report in the terms they use.

Indeed, the NASEM report's first recommendation is that "race should not be used as a proxy for human genetic variation. In particular, researchers should not assign genetic ancestry group labels to individuals or sets of individuals based on their race, whether self-identified or not."[6] However, I argue that *racial diversity* can be a proxy for genomic diversity. This may seem at odds with the NASEM recommendation. To the contrary, I see this chapter as addressing a pressing question and need that lingers over their recommendation. Race itself is not a proxy for the genotype of individuals because races are not biologically real. But we do need racial diversity in genomic research. I doubt the authors of the NASEM report would deny that. Thus, the inevitable question I raise here about how to square the call for diversity in genomic research with social constructionism about race needs addressing. I believe that clarity on this very limited way in which racial diversity can be helpful to genomic science and medicine, with extensive clarification on the limits of race's usefulness as a proxy, advances the same goals as the report. Race and racialized genetic ancestry themselves are not proxies for an individual's underlying genotype. But racial diversity in genomic research is needed to justly extend the putative benefits of genomic medicine to all.

In the first section of this chapter, I further discuss the importance of genetic diversity in genomic research and the call for racial diversity. In the second section, I explain the different conceptions of race: biological race realism, statistical race realism, and social constructionism. I explain why social constructionism is the most defensible conception of race and proceed through the rest of the chapter on the assumption that it is the correct view. Yet I will show how racial diversity in GWAS can be a proxy for genomic diversity, broadly speaking, even if race is a socially constructed category. In the third section, I use a novel analogy to do so. In the fourth section, I caution against the use of race as a proxy for individuals' genotypes in the clinical setting. Thus, even if race can be useful in promoting genomic diversity, its use as a proxy is quite limited and specific.

NEED FOR DIVERSITY IN GWAS

There is a need for genomic data that come from a diverse range of people. We cannot accurately extrapolate findings about gene variants and disease traits or drug responses from one population to another.

Populations are, roughly, interacting, interbreeding groups of individuals cooperating for survival.[7] Populations themselves are scientific constructs, not biological entities, that scientists posit for research purposes. Race and population are not interchangeable concepts. Change who you interact with, and you change your population. But this is not true of race.[8] Nonetheless, scientists frequently racialize populations, describing groups of people at the continental level because this level of grouping is familiar to and precedes population genetics. The definition of population does not preclude the possibility of interracial populations, obviously, but many genetic scientists construe populations along racialized lines in order to ensure roughly (what they think is) homogeneous ancestry among individuals within populations, an issue that will be discussed at greater length in the chapter by Carlos Andrés Barragán, Sivan Yair, and James Griesemer and in the chapter by Lisa Ikemoto. Predictions based on associations in one population may give rise to false positives in another population.[9]

This is for several reasons.[10] Allele frequencies vary among people by geography. When looking for medically relevant variants, geneticists compare those who exhibit the disease in question to controls who do not. If studies use people from different, geographically circumscribed populations, there is a risk of confounding alleles that vary among individuals due to a difference in ancestry with those that are associated with the disease in question. Controlling for population is meant to eliminate this confound. Additionally, a GWAS identifies *associations* between genes and traits, not the genetic *causes* of the trait. Given that alleles vary among populations, an identified marker of a trait may be linked to both common and rare alleles that cause the trait. The rare alleles may be frequent in some populations but not in others. Thus, a marker that is accurate in one population (where the rare allele is present) may give rise to a false positive in another population (where the allele is not present).

Another reason population diversity in genomic research is important is that scientists predict that rare variants (those that occur in less than 5 percent of the world population) will be more informative in predicting disease occurrence and drug response. Rare variants are often specific to populations.[11] These variants may be uncommon among people of European descent but present among other groups. Without genomic diversity, we are presumably failing to find many such rare variants.

This failure is especially acute because modern humans evolved in Africa. Some humans migrated out of Africa and populated the rest of the world. But these small, migrating groups carried with them only a subset of the genetic diversity that remained within Africa. Due to this genetic "bottleneck" and the fact that humans have been in Africa the longest, there is more genetic diversity among people with African ancestry than in other ancestral groups.[12] The exclusion of people from the African continent in genomic research poses an opportunity cost in identifying meaningful variants. For example, the discovery of *PCSK9* variants in people of recent African descent, which lower cholesterol in other ancestral groups, resulted in the successful development of the drug evolocumab, considered "the most important trial result of a cholesterol-lowering drug in over 20 years."[13] What other such discoveries are we failing to make for lack of diversity in genomic research?

This also signals a problem with geneticists' habit of using racialized population designations corresponding to continental-level populations. Given that there is much more genetic diversity in the "African population" than in other continental populations, lack of African diversity in genomic research may result in weaker associations between genetic markers and gene variants in African populations compared to European populations.[14]

The underrepresentation of certain non-white people in the data already entails a health-care disadvantage. Popejoy and Fullerton report that individuals of African and Asian ancestry—those often racialized as Black and Asian, respectively—more frequently receive nondefinitive test results or have variants of unknown significance.[15] Without racial and ethnic diversity in genomic research, those who are already underserved in the medical community—historically oppressed racial and ethnic minorities—will be further disadvantaged by a genomic medicine that does not include them.[16] For these reasons, many geneticists have called for racial diversification of GWAS to ensure the future potential benefits of genomic medicine apply to all.

WHY RACE IS NOT BIOLOGICALLY REAL

The need for genomic diversity, often construed as racial diversity, in genomic research may suggest to some that racial differences are reflected at the genomic level. The view that race has a genetic basis is a kind of biological race realism. Biological race realism is the view that race is a meaningful biological category that distinguishes differently racialized individuals on a biological or genetic level. Biological race realism is the conventional and perhaps common lay view of race. It is commonly held by scientists and physicians as well. In its original and crudest form, it is essentialist; it assumes that race is grounded in some biological essence—perhaps phenotype (e.g., physical features) or genotype (e.g., race-related genes)—that is inherited. In this view, race is discrete, meaning all of the people within one race share the essential features, while all of those outside the race lack these essential features. In this view, there are mixed-race people. But even this idea implies that there are "pure" racial groups that can then be blended, a widespread but mistaken belief that will be taken up in the chapter by Lisa Ikemoto.

This crude race realism has been widely dismissed by social scientists and philosophers. There are no biological features that comprise a discrete racial essence. Populations that correspond to the large continental groupings do not vary from one another in stark, discrete ways. Rather, phenotype and genotype among and within these large populations vary gradually—that is, *clinally*. We perceive there to be drastic and discrete morphological differences between groups of people (and thus infer discrete genetic differences) only when we compare individuals in (or with ancestors from) locations that are geographically distant from one another. If we look at people in (or with ancestors from) the places in between, we see gradual transitions in phenotype and genotype.

Crude race realism is obviously false. It is now being replaced by a *statistical race realism* in genetics. Some scientists and philosophers emphasize that, while there are no discrete populations corresponding to our common racial categories, there is structure to clinal genomic data.[17] The claim is that groups of individuals can be identified by genetic clustering among them—some individuals share distinctive groupings of genomic variants called haplotypes—that statistically correlates to having recent ancestry from particular geographic regions that are roughly continental. Perhaps these genetic clusters signify races.

But this new statistical view of race faces many criticisms. Some of the concerns are methodological. The clusters may be the artifacts of sampling strategies—for instance, using predefined populations that bias the data to produce racialized outputs; using small sample sizes for large, diverse geographic regions; preserving geographic distance between samples, which makes clinal differences look larger.[18] Further, generating statistically meaningful genetic clustering of populations that correspond to the familiar racialized continental-level groupings requires scientists to choose a number of clusters that reflects our race realist conception of the races.[19] The choice is arbitrary. Instruct the computer program to generate a large number, and you end up with 50 races, for instance, rather than the conventional 5. But this example shows that achieving an output that corresponds to our continental conception of race is possible only through human intervention in the data. In other words, these genomic clusters do not emerge from the data but are imposed on them.

There are many other concerns with statistical race realism.[20] But it will suffice to say here, even setting those important worries aside, that this statistical conception of *race* is far too revisionary to warrant the name. Races were originally theorized to be discrete, essentialist, and hierarchical. This new conception of race as continental genetic clusters is clinal, nonessentialist, and nonhierarchical. Shiao et al., who argue that these genetic clusters represent "clinal classes" homologous to race, see this departure of the race concept from its racialist roots as a defense of their argument. They say:

> Arguably, the origin of the essentialist criterion for biological differences lies less in actual science than in its use in the historical justifications for the categorical exclusion of nonwhites from political, economic, social, and cultural citizenship in the United States. By contrast, biological science does not require the white supremacist belief in species-level, much less greater, differences between human subspecies.[21]

But this is hardly a defense; it is a refutation. This view of race attempts to recuperate the old race realist categories for no scientifically motivated reason; the most benign reason is merely that these categories are familiar to us. Why adopt a term that is entirely inapt and loaded with a racist history to describe a novel putative biological phenomenon? A major worry is that calling this conception of human population structure "race" reifies race realism in the conventional understanding. This borrowed nomenclature facilitates the slide back into racialist thinking. Indeed, the move to a statistical conception of race is the continued social construction of race occurring in real time.[22] As I'll note in what follows, this same move happens in race-based medicine. In summary, statistical conceptions of populations are not races, and we shouldn't use race terminology to describe this position.

Race realism is in deep tension with the dominant academic view of race— one shared by many scientists and the majority of social scientists and humanities scholars—that our race concept is a social construction with no deep, meaningful biological reality.[23] Social constructionists argue that race is a socially constructed category for sorting human beings that has social reality and real effects on people's life prospects. In other words, races are real, but they are grounded in social facts, not biological ones. But this means that any biological or medical differences between the races have their source in historical processes or socioenvironmental causes, not mythical race-based genes.[24]

The social constructionist position about race is supported by the historical record, which traces race formation through time. Consider the changing racial categories used in the United States throughout its history.[25] In 1790 the US Census categorized people by their legal standing, using the categories of "Slaves," "Free White Females and Males," and "All Other Free Persons." By 1820 "Slaves and Free Colored Persons" were grouped together in one category in contrast to "Whites" and "Other Free Persons," illustrating the conflation (and true reality) of legal, political categories of hierarchy and race. By 1850 the first category became fully racialized as "Black and Mulatto" (and by 1890 included further fine-grained categorizations of "Black," "Mulatto," "Quadroon," and "Octoroon"), reflecting an entrenched rule of hypodescent where offspring of Black, white, or mixed parents inherit the political status of the parent deemed socially inferior. Only a political system, rather than rules of biology, could explain why children with a smaller portion of African ancestry are categorized with other Black people rather than with white people. This taxonomy functioned to keep the white race "pure." It limited the number of white people with full property and other civil rights. These rules are obviously socially and politically constructed and biologically arbitrary.

In 1860 the US Census added other racial designations—"Indian" and "Chinese"—alongside "Black/Mulatto" and "White," reflecting the contested legal status of Native Americans and Asian immigrants and their descendants as neither white nor Black. These categories morphed through time to the present-day mix of census race and ethnicity categories, always reflecting the social and political conditions and preoccupations of the day rather than biologically meaningful groupings.

The changes in the groups represented and rules about how people should classify themselves within them reflect social and political facts in the United States, including who could own land, changes in immigration demographics, and political solidarity. Regarding the last of these, the Asian American category came into existence in the 1960s as an explicit political rights movement echoing the civil rights achievements of the Black Power movement. Berkeley graduate students Yuji Ichioka and Emma Gee created the Asian American Political Alliance to unite discrete ethnic groups from Asia into one united political front, coining the term and hence race category *Asian American* at that time.[26] Another example is in the current racialization of Middle Eastern and North African people in the United States, many of whom, in the post-9/11 world, feel uneasy categorizing themselves as white and petitioned (unsuccessfully) to have the category *MENA* added to the US Census in 2020.[27] Social constructionism, not biological realism of any stripe, makes the most sense of these historically and socially grounded practices of race formation.

If races are socially constructed and not biologically real, then how can geneticists coherently advocate for racial diversity in order to achieve genomic diversity in research? Does the call for genomic diversity rely on the view that race categories are biologically real?

Before answering this question, it's worth noting that there are reasons to promote genomic research diversity that obviously do not assume race realism. One is that gene-environment interactions may differ by population—as a proxy for social circumstance.[28] Individuals, even with similar genes, in different environments may have different health outcomes. Including people with diverse backgrounds could eventually help scientists gain clarity on these interactions. This reason for genomic research inclusivity is not about genetic variation between racially defined groups but rather about socioenvironmental differences.

A COLORFUL ANALOGY

My aim is to show that increasing racial diversity among the genomic samples researchers use can increase genetic diversity in their research, even though race is not a biologically meaningful category. Specifically, what I'll argue is that diversity among socially constructed categories can be a proxy for diversity in some underlying physical reality, without the socially constructed categories being physically real. I endeavor to make the point through example, one that takes us away from the loaded debate about race. Take a natural phenomenon that is clinal—that is, gradual in variation—but upon which we've placed discrete, socially constructed categories. Variation in these socially constructed categories may function as a proxy in some limited ways for the natural variation in the clinal, physical phenomenon upon which they are imposed. The color spectrum, our conventions about color names, and the more precise wavelengths that produce different colors offer an apt example. To understand, we need to get nerdy about color for a moment.

Color is the perception of electromagnetic radiation in the visible spectrum of light.[29] Objects absorb some wavelengths of light while reflecting others. The wavelengths that are reflected back to an organism's eyes are, depending on the color receptors that organism has, perceived as a color. Humans can see wavelengths measuring 390–750 nanometers (nm).[30] For example, a blue object is one that reflects wavelengths measuring 450–495 nm. The spectrum of visible light, composed of wavelengths that produce colors when perceived by us, is gradual in nature. Indeed, the word "spectrum" has come to mean the organization of things that vary gradually in some regard that can be arranged from one extreme to the other. There are no discrete boundaries between areas on the spectrum; the wavelengths on the spectrum gradually change into one another, and so do the corresponding perceived colors.

We humans have given ranges within the light spectrum particular color names. We've roughly carved up the visible spectrum into color bands of red, orange, yellow, green, blue, and purple. But our color designations, which demarcate these bands into bounded, discrete groups, are socially constructed, and they vary by culture through place and time. For instance, the standard ROYGBIV seven-color partition of the spectrum, familiar to English speakers, originated with Aristotle, who theorized that there were seven colors just as there are seven musical notes.[31] Isaac Newton added orange and indigo to the already recognized colors of his time and place in order to achieve the Aristotelian ideal of seven colors and to honor the tradition of alchemy, in which the number seven has significance.[32] This is a vivid example of how cultural preference and human choice dictate the number of categories we impose upon a spectral reality. We are not carving the spectrum at its natural joints. Indeed, the spectrum—being clinal—has no such joints.

Consider other cultural variations. Greek and Russian speakers have distinct words for two different shades of blue, while others—for instance, speakers of English—use a broad category of blue for all hues within this range.[33] But the Tahitian, Tzeltal, and Japanese languages group blues and greens into one color category. In English, we consider red and pink distinct colors. Yet we have no such discrete separation between a saturated blue and a pastel one, the light-blue analogy to pink.

Ultimately, which colors we identify as distinct and how fine-grained our choices are may be the result of whether or not we have a purpose for making distinctions within broad color groups. In brief, color categories are culturally

VISIBLE SPECTRUM

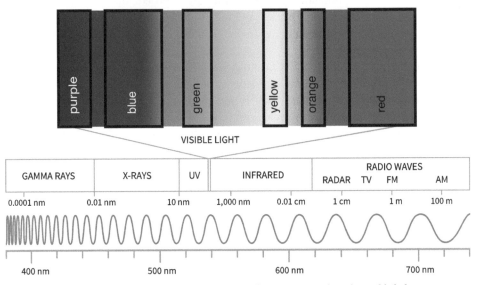

FIGURE 2.1. The conventional English-language colors are imposed on the visible light spectrum as discrete categories. Adobe Stock #229007362, modified by the author.

constructed. Colors are categories we impose on the spectrum. They are discrete, and while perhaps explainable by reference to culture, cultural development, or our physiology, they are arbitrary with regard to any distinguishing features of the light spectrum itself.

Yet our socially constructed color categories are still informative about the spectrum. Within each socially constructed color band is a group of wavelengths. For example, wavelengths of 620–750 nm produce the color red; those of 590–620 nm create orange. Let's say you are an eccentric collector: you collect electromagnetic wavelengths. You have many beautiful colored objects that reflect various wavelengths and produce lovely colors in the eye. But your collection is not very diverse. You have a lot of objects that reflect long wavelengths, the wavelengths that produce the color red. You have objects that are crimson, vermilion, shading into orange, and even some orangey-yellows. But you long to have a broader collection of wavelengths that represents the visible light spectrum of colors. If you wanted more wavelength diversity in your collection, you would do well to put out an advertisement for objects that are green, blue, and purple. Diversity in these broad color categories will be a good proxy for diversifying your collection with objects that reflect different wavelengths than do the objects in your current collection. Color categories, which are themselves socially constructed and do not map onto any physically real, discrete categories, can be a proxy for some underlying physical reality.

You can see the analogy. Genomic researchers have genomic samples mostly from individuals of proximate European ancestry. Many of these people will socially identify as white, as well. We know that allele frequencies among humans vary clinally by geography, with genetic distance correlating with geographic distance.[34] Thus, this narrow, geographically defined group represents only a limited set of human genomes. If you want to diversify your research genomically, people's ancestry can tell us which geographic region some or most of their proximate progenitors came from.[35] Race can be a rough guide to a person's proximate ancestry because it has been socially constructed to categorize people by visible traits that roughly correlate to having proximate ancestry from particular geographic regions. Thus, because of how race and ancestry are socially constructed, racial diversity can be a proxy for obtaining that geographically based genomic diversity. If you want more diversity in your mostly "white" genomic samples, you would do well to recruit for people of African, Asian, and Indigenous American ancestries, and so on. Race and ancestry of individuals, socially constructed categories, can be helpful, in this context, for indicating something biologically real. Namely, if you have more racial diversity on the whole among your samples, you should get more genetic diversity. But that does not mean that race and ancestry groupings are biologically real.

Like the color categories we've imposed upon the light spectrum, the race categories we've imposed upon geographical human populations are crude and arbitrary. We could have carved up the spectrum differently—for example, why not have a unique name for the yellow-orange of a marigold, why not carve up the blues into more discrete categories of aqua, cobalt, and periwinkle? Likewise, we could have carved up human populations differently. Why not, for instance, have more fine-grained categories for African populations, given all the genetic diversity in Africa?[36] But these categories can still do some work. Arbitrary though they are, we know that conventionally blue objects will reflect shorter wavelengths. Arbitrary though it is, we know that if genomic research focuses mostly on individuals who identify as white, it is lacking the genomic diversity that can be found in a more diverse sample of people who identify as Black, Pacific Islander, Asian, or Indigenous American.

So racial diversity can be a proxy for the purpose of getting more genetic diversity in our genomic research, just like color can help the fictional wavelength collector diversify their collection. But the fact that racial diversity is a proxy does not mean that race is biologically real, just as color diversity as a proxy for wavelength diversity does not mean that color categories map onto discrete features of the real light spectrum.

The point of the spectrum analogy is to simplify the issue at hand and put it in other terms in order to try to make sense of an otherwise novel and complicated phenomenon. But that simplification comes at a cost. The real pictures, for both real colors and the relation between race categories and genetic diversity, are far more complex. A simple analogy has its limits.

Complicating this analogy in accordance with reality, however, can be instructive. The spectrum itself is a simplification of color. We get pure colors represented on the spectrum, saturated colors like true yellows and greens. In real life, the color of objects is very rarely pure. Most real colors are mixes of the purer, more saturated colors, just as Mark Fedyk noted in his chapter that most gold in the world is alloyed. The color of a real apple is not vivid, saturated, pure red but rather is a brownish, grayish red. In reality, a real apple reflects back all of the spectrum wavelengths, just in different proportions, so that red wavelengths are dominant.

Something similar can be said of humans. Real humans are not representations of "pure" ancestral populations from which they came. There are no such things. Real humans are the products of complex human breeding histories. We all have genetic ancestors who came from many different places. Most of the alleles found in different frequencies in different parts of the world are present across the globe. The simplifying analogy of the color spectrum is inapt in at least one way to represent clinal human genetic diversity because any particular individual probably has genes that represent a crisscrossing, complex ancestral lineage that does not easily allow us to order individuals clinally along one dimension. Human breeding patterns and migration are dynamic; our ancestral populations did not stay in just one place, nor did they remain isolated from one another. Our genes reflect this dynamic, intermixing history.

Once we move to the more complex understanding of color, we can see the issue. A pure color spectrum can organize color linearly because it focuses on only one dimension of color: *hue*. Hue is the main local color of an object—for example, blue. But colors have two other main properties. *Saturation* is the purity of the color: is it a vivid, true blue with little else mixed in, or is it a desaturated, muted slate (a blue with gray in it)? *Value* is the depth of the color, how much black or white is mixed in: is it a dark navy or a pastel sky blue? Color theorists have endeavored for centuries to organize all the variation within colors in a way that could reflect these dimensions, coming up with complicated forms that relate all the colors in three dimensions. Move beyond the simple one-dimensional color spectrum, and this task proves quite difficult. Look at Munsell's color system.

Human genetic diversity is even more complicated. Humans have many gene variants, and although these vary clinally across global geographic distance, they vary "nonconcordantly," meaning that they do not covary together geographically.[37] If color diversity is difficult to represent with just three dimensions, organizing human genomic diversity with many more dimensions that can combine in multiple permutations is near impossible. So there is a limit to the spectrum analogy. In its simple form, it functions to show that diversity in socially constructed, discrete categories can serve as a proxy for diversity in some underlying, clinal physical reality. But more detailed, precise inferences from color to wavelength, or

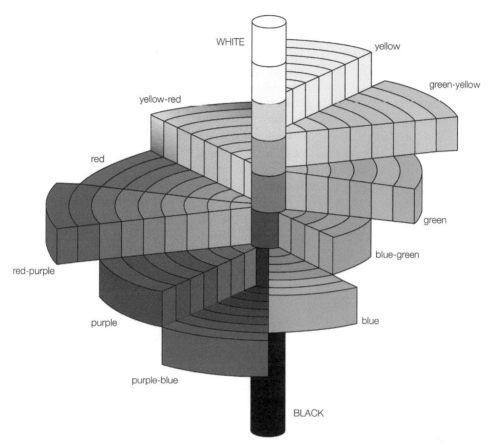

FIGURE 2.2. The Munsell color tree showing Albert H. Munsell's organization of colors by hue, value, and saturation, demonstrating the complexity of organizing spectral phenomena along more than one dimension of measurement. Universal Images Group North America LLC / Alamy Stock Photo.

race to genotype, are blocked when we add complexity to the model in accordance with messy reality.

RACE IN THE CLINICAL SETTING

We see the limit to the inference when we move race to the clinical setting. From the fact that racial diversity can serve as a proxy for genetic diversity among a large population of people, one might infer that the race of one individual can be a guide to their underlying genotype. But this a fallacious inference.

Back to our analogy. Say you are a comprehensive color collector, and you have objects representing the vast array of the pure spectrum. But you do not have any objects of the specific wavelength 578 nm. This wavelength falls into the green band on the spectrum. So you hoard a large set of green objects. Now, while this would be a better method for narrowing your search for 578 nm—say, as compared to scavenging for red objects—it is quite crude. You are not guaranteed to get 578 nm if you search for green. Green is quite broad a category to be a reliable proxy for something as specific as 578 nm. Importantly, if you have a particular green object in front of you, you cannot assume it is 578 nm.

The point strengthens when you consider the complexities we add to this analogy. Consider now that this is a real, colored object, not one merely representative of the pure color spectrum. You can't assume this green object doesn't have the wavelengths of other colors in it, since real green-colored objects have a complex mix of all the wavelengths. In fact, 578 nm could be present in any of your non–green-colored objects. In brief, with real colors, you can't infer from the presence of some color that you have either the presence or the absence of a particular wavelength.

Likewise, insofar as socially identified race is a rough proxy for people's genetic ancestry, then racial diversity can be a good proxy for getting more genomic diversity in your research.[38] But it won't guarantee you the presence or absence of any particular gene variants at the more fine-grained level, just as seeking green objects in no way guarantees that you will get the 578 nm wavelength. While certain gene variants are more frequent in certain global populations than in others, they may be present in lower frequencies all around. And within a population with a higher frequency of an allele, there will be some individuals who do not have it. Thus, focusing on an individual's race is not helpful in guessing which gene variants they may or may not have. One is not rationally licensed to move from the idea that ancestral background and race are proxies for genetic diversity within a sample with many people in it to inferring anything about the genes of an individual person in front of oneself from the way they look or how they identify.

Take the following example, in which race is used in the clinical setting as a proxy for the kinds of genes a person can have. Cystic fibrosis is a monogenic, recessive disease that primarily affects the lungs, resulting in excess production of mucus, difficulty breathing, lung infections, and hence shortened lifespan. It is commonly seen as a white disease, and one that affects Askhenazi Jewish people in particular. Dorothy Roberts tells the story of a two-year-old African American girl who presented in the emergency room with respiratory issues.[39] She had ongoing respiratory issues for years, until at age eight, a new doctor looked at her lung scan, not knowing her race, and accurately diagnosed her with cystic fibrosis. The child's race obscured the possibility of accurate diagnosis for her clinicians, who did not consider the possibility that people who were not socially identified as white could have cystic fibrosis. For this error, she went undiagnosed and untreated for a deadly lung disease for years. Race-based medicine runs this dangerous risk of licensing the assumption that a gene variant cannot be present in a person because of their race.

There are several reasons why race fails to be a good proxy in the clinical setting. First, our racial categories are too broad. Recall that there is more genetic diversity in the group of people with recent African ancestry than in any other continentally defined group. Recruiting people who self-identify as Black is a good way to cast a wide net for genetic diversity. But we cannot infer, with the appropriate level of accuracy, that any particular Black identified individual has a particular gene variant or trait. This may seem obvious. Yet the mistake is repeatedly made. Consider the claim that 40 percent of people of African ancestry are slower metabolizers of antidepressants, which is used as grounds in the clinical context for giving anyone of African ancestry—usually via self-report or clinician report of the patient as Black—a different dose than one would give a white person.[40] Yet according to this statistic, fewer than half of people of African ancestry have the trait. One using this racial heuristic in clinical treatment is undertreating more than half of their Black patients. And some non-Black patients may be overtreated since some percentage of them are presumably slow metabolizers of antidepressants.

Another reason race is a poor proxy in clinical practice—setting aside the point that racialized groups are too broad—is the arbitrariness of social rules for the assignment of race, which obscures the reality of "racial mixing." In the United States, for instance, someone who has half recent African ancestry and half recent European ancestry may identify as Black or African American due to the historical rule of hypodescent and the contemporary political understanding of racial group assignments. But this person has recent ancestry from at least two different, continentally defined ancestral groups. They may identify as Black in the clinical setting; they may be identified by the clinician as Black based on appearance and the historical, social rules for designating "mixed" ancestry. Both they and their clinician would be ignoring half of their ancestry if they are defined this way. These worries are especially sharp for diasporic Africans and Latinx individuals, given these groups' rich, diverse ancestry. Thus, even if ancestry is ultimately what matters and race is a crude proxy for ancestry, our social construction of race gives simple, typically discrete racial assignments to people with complex ancestral histories. This complexity is erased in the clinical encounter when a person self-reports their race or a doctor infers it based on their appearance. Racial determinations in the clinical setting are typically based on self-report.[41] Self-report of a politically created category is a very poor proxy for an individual's genetic profile.[42] Alternatively, a clinician surmising a patient's race based on their appearance, last name, or other features is not reliable either. None of these features is a reliable indicator of a person's complex ancestry. Further, although race is already a poor category for making these kinds of inferences, it is still less useful given the increased mixing of people who are differently raced. This is particularly concerning since the population of people who self-report being of two or more races is growing.[43]

Race is also a problematic category in the clinical setting because of its overtly social and political construction, which brings together people with diverse ancestry. Take, for example, the Asian race in the US context. As discussed earlier, "Asian American" is a political category intentionally created during the civil rights movement to unify people with ancestry from the Asian continent who are small minorities of the US population. Grouping together gave these subpopulations of Asia critical mass and relatively stronger political power in the United States. But *Asian*—derived from this conception of Asian American—in the US context includes people from South Asia, Southeast Asia, and East Asia, people who could not be grouped together as one homogeneous genetic population (nor would this be true for any of these constituent subcategories). Yet Asian is used as a racial category in medicine. For example, in the United States, spirometers, which measure lung function, are routinely "race-corrected" for Black and Asian patients.[44] I've already discussed the problem of using "Black" as a biologically meaningful category. Likewise, what started as a category for an explicit political power movement among Asian Americans has been biologized by the medical establishment as an indicator of innate biological response in the clinical setting. There is no scientific warrant for this.

The use of race in the clinical setting suggests that "racially profiling doctors" have internalized crude race realism in making their assumptions about patients. Were crude race realism true, it would better allow the inference from individual to gene or trait because race realism is the view that races are discrete and essentialist. So being of race X means having the features that people of race X have. This kind of race realism is false. We have no justification for sliding back into it in medical practice. In addition to the dangers of misdiagnosis, this practice sends the message that crude racialist races are real.

But the statistical notion of race, endorsed by some scientists and doctors, does not license the inference either. At best, among a group of people similarly racialized, we see an increase in some clinically relevant alleles in the group. But one is guilty of committing the ecological fallacy when one moves from this group-level statistic to inference about individual risk. Higher incidence of Y among a defined population does not mean an individual member of the group has a higher risk of Y. This is starkly the case when the criterion for grouping itself is not medically or biologically meaningful.

One might interject and claim that these alleged statistical associations between race and certain gene variants could be meaningfully deployed in medicine, even if they are not perfect. But that is too hasty. Determining whether race is a good proxy for genes requires settling a value-based assumption. Our tolerance for a proxy's accuracy can vary from context to context. The context tells us how sensitive and specific the proxy must be for our purposes. A highly sensitive test gives us a high rate of true positives. A highly specific test gives us a high rate of true negatives. We must evaluate the risks if our proxy is not precise and weigh them against the benefits of having a proxy with its particular level of accuracy. Context tells us what risk tolerance we should have.

For instance, race may be an appropriate proxy for genomic research recruitment. In recruiting for genomic research, all we need to do is cast a very wide net in order to include many people from many different geographical areas. The risks to individuals in doing so are relatively minimal.[45] We are including them in a genomic sample database but not otherwise interacting with them. Yet, in the clinical setting, the lack of precision in inferring genes from race has real, tangible risks, including misdiagnosis and inaccurate drug dosing. The case of the Black child with cystic fibrosis is instructive. The risks of misdiagnosis using race as a proxy for genotype are high—they are life or death. In this context (and for the other reasons already mentioned), race is not a good predictor of genotype. This then limits the use of race as a proxy for genomic diversity in genomic medicine.

CONCLUSION: RACIAL DIVERSITY CAN BE A LIMITED PROXY FOR GENOMIC DIVERSITY

Human genetic diversity changes gradually across the globe, with genetic differences correlating with geographical distance. Since the US race construct is based on taxonomizing people by relatively recent geographic place of origin for proximate ancestors, this common race construct is a rough proxy for large, continental ancestral place of origin. For this reason, race might be a good proxy for casting a wide net to increase genetic diversity in genomic research. For reasons of justice, we should care about and advocate for racial diversity in genomic research. All people should be able to benefit from the medical findings of these studies. But advocating for racial diversity in genomic research does not require a commitment to biological race realism. Racial diversity, relying on race as a socially constructed category, can serve as a proxy for genomic diversity; more racially diverse people included among genomic samples should correlate with more genomic diversity. But we should also be very aware of the limitations of this relationship. We'll need racial diversity to develop genome-based, precision medicine equitably; but race should not serve as a proxy for making genetic inferences about individuals in the clinical setting.

There is another way that race can be relevant to health outcomes, and that is through socially and environmentally mediated processes. Racism, differential access to health care, exposure to pollutants, and so on have deep and lasting health outcomes and are differentially distributed by race. My hope is that the preoccupation with the relationship between genes and race does not obscure this more promising avenue for understanding racial disparities in health.

NOTES

1. "GWAS Diversity Monitor," accessed August 24, 2023, https://gwasdiversitymonitor.com.

2. Carlos D. Bustamante, Esteban Gonzalez Burchard, and Francisco M. De La Vega, "Genomics for the World," *Nature* 475, no. 7355 (2011): 163–65, https://www.ncbi.nlm.nih.gov/pmc/articles/PMC3708540/; Alice Popejoy and Stephanie M. Fullerton, "Genomics Is Failing on Diversity," *Nature* 538, no. 7624 (2016): 161–64, https://www.ncbi.nlm.nih.gov/pmc/articles/PMC5089703/; Jonas Korlach, "We Need More Diversity in Our Genomic Databases," *Scientific American: Voices*, December 4, 2018, https://blogs.scientificamerican.com/voices/we-need-more-diversity-in-our-genomic-databases/.

3. See Michael Montoya, *Making the Mexican Diabetic: Race, Science, and the Genetics of Inequality* (Oakland: University of California Press, 2011); Dorothy Roberts, "Debating the Cause of Health Disparities: Implications for Bioethics and Racial Equality," *Cambridge Quarterly of Healthcare Ethics* 21 (2012): 332–41; Jay S. Kaufman, Lena Dolman, Dinela Rushani, and Richard S. Cooper, "The Contribution of Genomic Research to Explaining Racial Disparities in Cardiovascular Disease: A Systematic Review," *American Journal of Epidemiology* 181, no. 7 (2015): 464–72.

4. Hispanic/Latinx people, often taken to be a race by Americans, are considered an ethnicity in this taxonomy, although the contemporary racialization of this group is yet one more example of the social construction of race in real time.

5. National Academies of Sciences, Engineering, and Medicine (NASEM), *Using Population Descriptors in Genetics and Genomics Research: A New Framework for an Evolving Field* (Washington, DC: National Academies Press, 2023), https://nap.nationalacademies.org/catalog/26902/using-population-descriptors-in-genetics-and-genomics-research-a-new.

6. NASEM, *Using Population Descriptors*, 7.

7. Roberta Millstein, "Thinking about Populations and Races in Time," *Studies in History and Philosophy of Biological and Biomedical Sciences* 52 (2015): 5–11. Geneticists are not consistent in their definition and use of the term *population*, however.

8. Joshua Glasgow, "Is Race an Illusion or a (Very Basic) Reality?," in *What Is Race? Four Philosophical Views*, by Joshua Glasgow, Sally Haslanger, Chike Jeffers, and Quayshawn Spencer (New York: Oxford University Press, 2019), 111–49.

9. Arjun K. Manrai et al., "Genetic Misdiagnoses and the Potential for Health Disparities," *New England Journal of Medicine* 375 (2016): 655–65, https://www.nejm.org/doi/full/10.1056/NEJMsa1507092; Popejoy and Fullerton, "Genomics Is Failing on Diversity."

10. For claims in this paragraph, see Bustamante, "Genomics for the World."

11. Bustamante, "Genomics for the World."

12. Bustamante, "Genomics for the World"; Serena Tucci and Joshua M. Akey, "The Long Walk to African Genomics," *Genome Biology* 20 (2019): 1–3. Humans also migrated back to Africa, increasing genetic diversity in Africa in yet another way.

13. James Gallagher, "'Huge Advance' in Fighting World's Biggest Killer," *BBC*, March 17, 2017, https://www.bbc.co.uk/news/health-39305640; George Adigbli, "Race, Science and (Im)precision Medicine," *Nature Medicine* 26 (2020): 1675–76, https://www.nature.com/articles/s41591-020-1115-x.

14. Bustamante, "Genomics for the World."

15. Popejoy and Fullerton, "Genomics Is Failing on Diversity." This finding is for whole genome and exome sequencing, which has slightly better but still inequitable diversity among its sample participants compared to GWAS, which scans the genomes for select variants.

16. Bustamante, "Genomics for the World."

17. The genetic clustering data is from Noah A. Rosenberg et al., "Genetic Structure of Human Populations," *Science* 298, no. 5602 (2002): 2381–85. Notably, Rosenberg et al. do not call these clusters races, writing in a follow-up publication: "Our evidence for clustering should not be taken as evidence of our support for any particular concept of 'biological race'"; Rosenberg et al., "Clines, Clusters, and the Effect of Study Design on the Inference of Human Population Structure," *PLoS Genetics* 1, no. 6

(2005): e70, 668. In contrast, the following authors think these clusters are evidence of biological races. See Jiannbin Lee Shiao, Thomas Bode, Amber Beyer, and Daniel Selvig, "The Genomic Challenge to the Social Construction of Race," *Sociological Theory* 30, no. 2 (2012): 67–88; Quayshawn Spencer, "How to Be a Biological Racial Realist," in *What Is Race? Four Philosophical Views*, by Joshua Glasgow, Sally Haslanger, Chike Jeffers, and Quayshawn Spencer (New York: Oxford University Press, 2019), 73–110.

18. Joan H. Fujimura, Deborah A. Bolnick, Ramya Rajagopalan, Jay S. Kaufman, Richard C. Lewontin, Troy Duster, Pilar Ossorio, and Jonathan Marks, "Clines without Classes: How to Make Sense of Human Variation," *Sociological Theory* 32, no. 3 (2014): 208–27; Ann Morning, "Does Genomics Challenge the Social Construction of Race?," *Sociological Theory* 32, no. 3 (2014): 189–207.

19. Fujimura et al., "Clines without Classes"; Michael James, "Race," *Stanford Encyclopedia of Philosophy* (2020), https://plato.stanford.edu/entries/race/.

20. Fujimura et al., "Clines without Classes"; Morning, "Does Genomics Challenge the Social Construction of Race?"

21. Shiao et al., "Genomic Challenge to the Social Construction of Race," 70. This false separation of science from systems of power and politics is alarming and inaccurate.

22. See Morning, "Does Genomics Challenge the Social Construction of Race?," 90.

23. For a detailed discussion of social constructionism in both its political and its cultural varieties, see Sally Haslanger, "Tracing the Sociopolitical Reality of Race," and Chike Jeffers, "Cultural Constructionism," both in Joshua Glasgow, Sally Haslanger, Chike Jeffers, and Quayshawn Spencer, *What Is Race? Four Philosophical Views* (New York: Oxford University Press, 2019).

24. Among philosophers, there is another camp. *Eliminativists* about race argue that since race is a biological concept and there is no meaningful biological reality to race, we should eliminate our race concepts from scientific (and other) discourse entirely. Instead, we should speak of *racialized groupings*, which are real, but explicitly socially constructed. The differences between eliminativists and social constructionists need not concern us here. The eliminativist can substitute claims in this paper about race as claims about racialized groupings. For more on eliminativism, see Glasgow, "Is Race an Illusion or a (Very Basic) Reality?"

25. "Measuring Race and Ethnicity across the Decades: 1790–2010," US Census Bureau, accessed August 24, 2023, https://www.census.gov/data-tools/demo/race/MREAD_1790_2010.html; Anna Brown, "The Changing Categories the U.S. Census Has Used to Measure Race," *Pew Research Center*, 2020, accessed February 8, 2024, https://www.pewresearch.org/fact-tank/2020/02/25/the-changing-categories-the-u-s-has-used-to-measure-race.

26. Anna Purna Krishnamurthy, "In 1968, These Activists Coined the Term 'Asian-American'—and Helped Shape Decades of Advocacy," *Time*, May 22, 2020, https://time.com/5837805/asian-american-history/.

27. Neda Maghbouleh, Ariela Schachter, and René D. Flores, "Middle Eastern and North African Americans May Not Be Perceived, Nor Perceive Themselves, to Be White," *Proceedings of the National Academy of Sciences* 119, no. 7 (2020): e2117940119.

28. Bustamante, "Genomics for the World."

29. There is a large literature in philosophy on the nature of color and color perception that I am setting aside in order to offer an accessible analogy for my purposes here.

30. Adam Rogers, *Full Spectrum: How the Science of Color Made Us Modern* (New York: Houghton Mifflin Harcourt, 2021), 2.

31. Rogers, *Full Spectrum*, 38.

32. Rogers, *Full Spectrum*, 53–54.

33. Rogers, *Full Spectrum*, 151.

34. John H. Relethford, "Biological Anthropology, Population Genetics, and Race," in *The Oxford Handbook of Philosophy and Race*, ed. Naomi Zack (New York: Oxford University Press, 2017), 160–69, 163.

35. Ancestry itself is socially constructed. We all have ancestors in every generation of humans. Choosing as the relevant ancestry ones that pertain to the continental geographic regions is both a spatial choice

(which regions, how large or small?) and a temporal one (a particular snapshot in time where the relevant groups can be tied, roughly, to specific locations). These choices are informed by our socially constructed conception of what the races are.

36. Geneticists do this to some extent, specifying populations within Africa. They also frequently lapse into using the broader continental groupings.

37. Fujimura et al., "Clines without Classes."

38. Although we each have many ancestries, not one.

39. Dorothy Roberts, "What's Race Got to Do with Medicine?," *TED Radio Hour*, NPR, February 10, 2017, https://npr.org/transcripts/514150399.

40. Sally Satel, "I Am a Racially Profiling Doctor," *New York Times*, May 5, 2002, https://www .nytimes.com/2002/05/05/magazine/i-am-a-racially-profiling-doctor.html. Satel is a publicly vocal proponent of what she calls "racial profiling in medicine," and for that reason she may be seen as representing an extremist view. But her thinking is representative of a large swath of clinicians and, more generally, medical practices that use race as a proxy in the clinical setting. For a recent assessment of race-based medical practices, see Jessica P. Cerdeña, Emmanuella Ngozi Asabor, Marie V. Plaisime, and Rachel R. Hardeman, "Race-Based Medicine in the Point-of-Care Clinical Resource UpToDate: A Systematic Content Analysis," *eClinicalMedicine* 52 (2022), https://thelancet.com/journals/elinm/article /PIIS2589-5370(22)00311-X/fulltext. For a discussion and criticism of race-based clinical algorithms, see A. Vyas, Leo G. Eisenstein, and Davis S. Jones, "Hidden in Plain Sight—Reconsidering the Use of Race Correction in Clinical Algorithms," *New England Journal of Medicine* 383 (2020): 874–82. They state: "Most race corrections implicitly, if not explicitly, operate on the assumption that genetic difference tracks reliably with race."

41. Youssef Roman, "Race and Precision Medicine: Is It Time for an Upgrade?," *Pharmacogenomics Journal* 19 (2019): 1–4, https://www.nature.com/articles/s41397-018-0046-0.

42. Michael Root, "Race in the Biomedical Sciences," in *The Oxford Handbook of Philosophy and Race*, ed. Naomi Zack (New York: Oxford University Press, 2017), 463–73.

43. Roman, "Race and Precision Medicine."

44. Lundy Braun, "Race, Ethnicity, and Lung Function: A Brief History," *Canadian Journal of Respiratory Therapy* 51, no. 4 (2015): 99–101.

45. That's not to say there are no ethical issues for GWAS. See Stephen J. O'Brien, "Stewardship of Human Biospecimens, DNA, Genotype, and Clinical Data in the GWAS Era," *Annual Review of Genomics and Human Genetics* 10 (2009): 193–209, https://doi.org/10.1146/annurev-genom-082908 -150133l; Jantina de Vries, Susan J. Bull, Ogobara Doumbo, Muntaser Ibrahim, Odile Mercereau-Puijalon, Dominic Kwiatkowski, and Michael Parker, "Ethical Issues in Human Genomics Research in Developing Countries," *BMC Medical Ethics* 12 (2011): 5, https://doi.org/10.1186/1472-6939-12-5.

3

Eventualizing Human Diversity Dynamics

Admixture Modeling through Time and Space

Carlos Andrés Barragán, Sivan Yair, and James Griesemer

Since the introduction in the early 2000s of direct-to-consumer genomic ancestry (DTC) testing, human genomic knowledge has increasingly challenged popular imagination about what human diversity is—its past, present, and future.[1] Set in motion by academic research projects and by many small, medium, and large private companies, DTC has become a multibillion-dollar industry that prominently targets a sociocultural curiosity about DNA,[2] human origins, and a desire to have them narrowed down to the individual level. As producers and consumers of human genetic and genomic knowledge, we can be hopeful regarding what this type of knowledge can offer about our history and future as a species. Yet, optimism can be mistaken with a misguided sense that DNA data can exhaustively and unequivocally answer who we are by situating our individual histories within the history of *Homo sapiens*.[3] The blooming of DTC was possible through research work on human population genomic ancestry studies (HPGA). Both HPGA and DTC (illustrated by research enterprises such as the Genographic Project and companies such as Ancestry and 23andMe) have greatly complicated how actors and audiences (with multiple backgrounds and motivations) think and argue about complex and ambiguous concepts such as ancestry, ethnicity, history, identity (individual and/or collective), and/or race.[4]

Academic and public exchanges between life scientists, consumers, sample donors, bioethicists, journalists, lawyers, legislators, public servants, social scientists, and others around these concepts remain problematic due to the ways genomic knowledge is produced, disseminated, and consumed. By this we mean that *production* conditions—e.g., the conceptual, technical, and inference apparatuses used to sequence and interpret genomic data—are not necessarily made explicit or understandable in research articles or genomic services because they are supposed to circulate mainly among specialists in HPGA, for whom such details constitute common background knowledge.[5] In the context of DTC products (tests, datasets, platforms, and narratives), the technological and statistical complexity behind linking individuals in time and space into accounts of ancestry is reduced to an oversimplified, and most of the time anachronistic, use of discrete ancestry-related categories, such as ethnic denominations, geographic locations, and/or nationalities. In the case of the *dissemination* of human genomic knowledge, journalists' insights can be limited by the characteristic brevity of their work or by sensationalist media approaches that contribute to the reification of the categories mentioned above. In the context of human genomic knowledge *consumption*, DTC producers and research participants/consumers are not necessarily interested in questioning or dissecting its "secret sauce," as human geneticist Spencer Wells described the theoretical and methodological scaffolding used to build and articulate narratives of "deep ancestry" in the context of the Genographic Project.[6] Doing so can compromise the overall perception of how robust these products and their findings are *vis-à-vis* the theoretical and statistical assumptions that HPGA studies require to render and interpret genomic datasets.

Perhaps one of the most widely shared assumptions regards the very concept of "ancestry." The fact that there is no effort to define ancestry in most contexts where it is being discussed turns paradoxical since, in the case of both academic scientific research (HPGA) and DTC, findings and products are supposed to teach multiple audiences more about it. In the last decade, several social scientists have pointed out how the concept is taken for granted and what the consequences of that are for multiple vulnerable communities as they consume and/or contest genomic knowledge.[7] Life scientists have pointed to similar caveats while being explicit about the limitations of the insight that genomic data and interpretative tools can offer.[8] There are also examples of collaboration between life and social scientists to deliver critical insights and criticisms.[9] The importance of ancestry is enhanced by the fact that its potential meanings are a base for individual and collective identity and the concept is crosscut by other polysemic concepts, such as ethnicity and race. Yet ancestry is not the only concept around which misunderstandings emerge.

There are other concepts that are perhaps less ubiquitous but that intersect and articulate with the concepts just mentioned in key ways. *Genomic admixture* is one of them. In a very general way, the concept captures a process through which human individuals from populations that have been separated for a long time breed together and produce offspring whose genomic lineages trace back to both populations. This concept led to the production of different technologies (methods) to track the frequency of disease-causing genetic variants by linking them to ancestral populations for contemporary "recently admixed" populations, such as African Americans, Mexican Americans, or Latinos. One of the names given to such applications in human genomics has been *admixture mapping*. Since the 2000s, a shared goal in admixture mapping studies has been the construction of ancestry informative markers (AIM) for different ancestral populations (e.g., African, European, and Native American populations in the sixteenth century and their descendants in the Americas). The logic and the theoretical framework behind admixture are that the frequency of genomic variants thought to cause a disease might be higher in ancestral populations known for having a higher incidence of a given disease than in other ancestral populations not known for as high an incidence of such disease.[10] Although admixture studies have generally been framed and regarded as positive contributions to understanding human evolution,[11] the biological basis of diseases, and the socialization of DTC, some life scientists themselves have also critically emphasized the need to avoid deterministic interpretations of such genomic factors and datasets to think about ancestry and biomedical risk—something that general audiences in industrialized countries think about in equal terms and that helps to explain the popularity of DTC tests.[12]

The current body of literature on human admixture studies is enormous and includes applications that re-situated HPGA data and insights into areas such as biomedicine and forensics (respectively, HPGB and HPGF), whose analysis requires far more work than we can report and reflect on for this volume.[13]

In this chapter, we offer insights on how the concept of genomic *admixture* is currently being used in HPGA studies and how such uses open up spaces for misunderstandings as producers and consumers of genomic data re-situate findings to think and talk about ancestry and identity through time and space.[14] Within HPGA studies, we further narrow our scope by focusing on recent genomic research studying the peopling process of the Americas. Our aim is not to produce an exhaustive literature review of every single published article reporting on these migratory processes. The analysis provided here is a first step toward the tracking of research studies that aim to identify broad-scale (spatial and temporal) patterns of movement/migration/interbreeding. This will allow us to document and analyze key aspects of how the concept of admixture is understood and deployed to explore models of how humans populated the Americas—a process that is argued to have started 20,000–15,000 years before the present (BP). In this context, we ask how concepts and assumptions of admixture guide research design processes (e.g., modeling practices, assumptions behind choices for sampling strategies, and/or methods and tools deployed). Likewise, we track what assumptions about ancestry and identity are at play as other audiences re-situate findings that involve narratives about population admixture through time and space. These goals are part of a larger research agenda to understand how scientific knowledge is being re-situated between settings (e.g., laboratories, research institutes, companies) and/or between audiences with several degrees of expertise.[15] In this context, we understand scientific knowledge as a complex assemblage of objects that includes—but is not limited to—research questions, models, datasets, findings, visualizations, narratives, etc.

Our argument is that, as different actors re-situate *admixture* findings in and from HPGA, this *re-situation* complicates and disrupts the necessary contexts needed to evaluate the robustness of the genomic knowledge that is supposed to be generated in tandem with other scientific objects. This is the case whether admixture is set to travel as a genomic concept, or as an object that can be used to assess discrete or continuous states of genomic ancestry for individuals (and the aggregates they represent), or as a synonym of polysemic popular ideas of racial and ethnic mixture across the globe. In the form of questions: How well are genomic admixture data and findings traveling?[16] What sort of misunderstandings can happen if the concept is re-situated without also re-situating other objects (e.g., metadata about the populations being represented and sampled), or without necessary clarification about the assumptions it needs to be useful in a given workflow (e.g., in characterizing how homogeneous or heterogeneous a population can/must be as it changes through time and space)? We argue that, when admixture is set to travel and is re-situated, misrepresentations and misunderstandings can take place when producing HPGA and DTC work and critical work about them. Although there is no single model to predict how re-situations beyond the limits of specialists (i.e., life scientists) would take place, we contend that life scientists do have a vantage point to minimize potential misinterpretations of the concept as a euphemism for widespread colonial concepts of racial mixture, regionally illustrated by a plethora of other concepts, such as *mestizaje, métissage, mestiçagem,* and miscegenation.[17] This will require, for example, making explicit how the concept is understood in life scientists' grants and publications, how it is linked to the design of the research workflow (e.g., population representation, sampling, and metadata production), how it articulates the production of findings, and what the potential pitfalls are if the geographic and temporal scopes of analysis are challenged by other audiences.

What follows is a brief description of our analysis for the subsequent sections of this chapter. In the second section, we offer a genealogy of what admixture, as a genetic concept, is supposed to capture about human populations. We also make an argument about how its current uses afford different levels of abstraction during modeling processes and the enrollment of multiple assumptions to make them happen. In the third section, we describe and analyze published research outcomes that have used contemporary and ancient human DNA samples to study, complement, and contest research questions, models, data, and findings about the peopling of the Americas that in the past were mostly a specific domain of archaeologists, biological anthropologists, and paleontologists. In the fourth section we analyze how life scientists themselves and other key actors (such as journalists) re-situate genomic admixture findings and what consequences such processes can have in terms of robustness. Finally, in the fifth section, we discuss some strategies that can be used by life scientists to minimize potential misunderstandings as their models, concepts, and findings are re-situated by other colleagues, actors, and large audiences.

THINKING WITH A CONCEPT: GENOMIC ADMIXTURE

Admixture: The formation of a hybrid population through the mixing of two ancestral populations.[18]

The brevity of the definition of admixture offered by life scientist Mark Jobling and colleagues in their popular textbook *Human Evolutionary Genetics*—used for teaching undergraduate and graduate students in Anglo-Saxon contexts—stands as a stark contrast to the complexities of the genomic phenomenon it aims to capture.[19] The abstraction embedded in the concept starts with the circumscription of two genetically distinct populations (through time and space) and the emergence of a third (or more) population(s) through breeding. Semantically, the Latin root *admixtus* gives both the verb (*admix*) and noun (*admixture*) forms in the English language, which go back in time as far as the fifteenth century and mean the blending of two or more different things into a new one.[20] These are semantic dimensions that have older and wider historical and sociocultural contexts beyond the appropriation of the term in human population genetics since the 1960s.

Accumulated archaeological, paleontological, historical, and human population genetics research outcomes have shown that the peopling of the planet was possible due to complex processes of migration, settling, and further migration.[21] What the admixture concept adds to the recent genomic study of such large-scale processes is the ability to explore different models to track migration and population interactions that have contributed to shaping human genetic diversity—understood as the total amount of variation in genes or whole genomes of individuals within or among population(s)—and its structure. The application of such models becomes trickier as the elements of an admixture event (two isolated populations and a new one with multiple ancestries) require further characterization, usually offered in terms of genetic ancestry profiles that can have multiple population sources,[22] depending on how far in time and how wide in space such profiles or "diversity panels" are designed to go.

This is a good point at which to emphasize that such *genetic* or *genomic ancestry*—that is, the sources of genomic material within a genome (represented by a living tissue donor or by an ancient bioarchaeological specimen, for example)—is different from other characterizations of connections between individuals and the populations they could represent, such as *genealogical ancestry* or concepts such as *genetic similarity*. Geneticists Mathieson and Scally have emphasized the need to undo the conflation of these concepts, an outcome of the ubiquitous narratives set in motion by DTC, in order to avoid the oversimplification and misinterpretation of genomic data turned into ancestry substantiations.[23] In the case of *genealogical ancestry*, the relationship created is between an individual and ancestors in their family tree and characterizations of interests such as nationality or surnames. Rather than referring only to an individual's pedigree, *genetic ancestry* is a subset of a family tree through which geneticists track genetic material that is inherited by an individual. On the other hand, the concept of *genetic similarity* between individuals (and the populations they could represent) is better understood as a "summary" of genetic variation built from multiple past or current individuals to represent specific populations (e.g., ancestral or admixed). This is a process that is susceptible to multiple contingencies that include explicit and implicit biases when deploying data about data—metadata—to design and execute research workflows (e.g., the naming, sampling, and representation of existing, deceased, or unknown populations). In the context of DTC, such summaries substantiate narratives of individual identity by conflating genetic and genealogical ancestry data in problematic ways. The most evident challenge is the assumption of population genomic continuity when making statements about "African," "European," or "Amerindian" ancestry that won't hold meaning when focusing on a finer scale (e.g., smaller than a continent) or for time periods that lack historical records.

These subtleties matter because, if a consumer of a genetic ancestry test uses its results to describe or corroborate their preconceived ancestry—individual, cultural—as *admixed*, such characterization may be the result of assumptions about genetic similarity to present-day individuals, rather than an account of ancient or past key admixture population events developed to track and understand the spread of *Homo sapiens* around the world. The latter is the kind of inference that matters most to some life scientists in HPGA interested in the peopling of the planet. Yet these conceptual distinctions in how genomic knowledge is being re-situated from HPGA to DTC contexts are not the only interesting challenges requiring some epistemological considerations when focusing on the concept of genetic admixture and admixture mapping as an application of ancestry identification.

Again, the most basic unit of a theoretical admixture event requires modeling with two geographically isolated populations that breed and produce a new admixed one that should reflect ancestors from multiple sources. However, retrospectively, it follows logic to assume that the two isolated populations were at some point (earlier in time) likely admixed from older populations. Likewise, prospectively, it is possible that the third population could become in the future an isolated one (geographically) and potentially an ancestral one in different admixture events. These aspects of the model prompt several questions. On one hand, how much time does it require for a population to become isolated enough to contribute to a new one through recombination? Is this something that is best estimated in terms of years, or in terms of generations? On the other hand, for how long does an interbreeding process between two isolated populations need to go on before it can be called an admixture *event*? Furthermore, do different admixture events show lower or higher admixture levels, or are they simply different depending on the distribution of variants that amount to genomic diversity given how these are used to produce diversity panels? We are not interested in arguing that these types of considerations are unknown to life scientists when genomic admixture models are re-situated between labs and research programs.[24] However, its discussion is not so prominent in the training of new generations of researchers (graduate level) and is not necessarily made explicit in the dissemination of admixture mapping research findings in scientific research journals, which usually strive for brevity in content. Although command over these subtleties is achieved through practice (i.e., senior researchers who have accumulated theoretical, statistical, and computational modeling experience), our point is that this type of modeling subtlety and the larger set of scaffolded assumptions necessary to model must not only be made explicit but also contextualized in the process of data interpretation, for scientists and the public alike.

Modeling assumptions in HPGA in general or in admixture mapping in particular do not necessarily signify flaws in the scientific knowledge being produced. Assumptions can also be understood as key scientific objects that facilitate the production of models and should be appraised as the research workflow takes place. Unaddressed assumptions are the concerning instance, since they can lead to overinterpretation of datasets, which has been most evident so far in DTC. As a fairly recent area of research, practitioners of admixture mapping in HPGA are currently debating how its insights can be both descriptive and predictive and are thus setting research priorities. Our premise is that robustness can be built only by addressing the extent to which genomic data can be forced to speak about recent and ancient population admixture events.

The earliest challenge for the application of a basic model of admixture started when researchers pondered the identification of the best possible DNA donors to produce admixture studies. In abstract terms, such a scenario would require samples from the two isolated populations and from the new admixed one. However, that scenario is almost impossible to encounter when studying concrete human populations. When HPGA researchers started networking in the 1980s to produce large-scale research projects to answer questions about the peopling of the world (e.g., the Human Genome Diversity Project), one of the most basic consensuses reached was that the priority was to collect tissue from living populations that had managed to stay relatively isolated from global demographic processes set in motion by European kingdoms and their colonial enterprises.[25] Back then, the idea of sampling individuals representing genetically admixed populations, described as "melting pots," was anything but appropriate for the reconstruction of the history of human populations predating historical repositories. It took multiple technical and scientific developments, and no small number of debates among life scientists, to value concepts such as *admixture linkage disequilibrium* (ALD)[26] as potential strategies for characterizing ancestral populations.[27] ALD led to a method known as *mapping by admixture linkage disequilibrium,* or MALD,[28] whose logic was substantiated by the association between an allele and a trait (marker) for the purpose of assigning gene(s) to a linkage group. A few years later, in 1998, another method was proposed that left aside the linkage disequilibrium between alleles and a trait and focused on the association between a local chromosomal ancestry and a trait.[29] This method was coined *admixture mapping*[30] and has been used to describe the burgeoning of a research program that applies ancestry identification to learn about the genetic basis of phenotypic variation (e.g., diseases) and to yield the consolidations of AIM panels for different "admixed" populations.[31]

The corpus of research produced during the 2000s and 2010s suggests that ancestry patterns found in so-called or historically self-identified admixed populations are useful for understanding larger evolutionary aspects of human evolution (e.g., timescales, mechanisms), whereas scientists had previously believed that only isolated populations (i.e., ethnic minorities across the globe) could serve this purpose.[32] Yet the interest in sampling ethnic minorities as a means to understand "deep" questions about humanity's journey didn't decrease in the past two decades. Its importance has actually been enhanced by the emergence of ancient DNA (aDNA) as a new "record" or "archive" and by the establishment of paleogenomics[33] as a research area in its own right for "rewriting" human evolutionary history.

In the case of the peopling of the world, biological anthropologists, archaeologists, paleontologists, and now human geneticists have been building models and extensive datasets over the past two decades to reconstruct and characterize the long migration journeys from what we today call Africa (100,000–60,000 BP) to the very end of South America (15,000–10,000 BP). Despite how new datasets complement and/or challenge previous findings, we can point out that a conceptual and methodological constant across these disciplines is the use of geographic areas (e.g., continents, regions, localities) as units of analysis to represent specific population(s) and thus to articulate specific admixture event(s). The point is that, in reality, such units of analysis could represent *multiple* populations and multiple admixture events (interpreted as ancestries). At stake is how, in order to narrate and visualize larger patterns of movements of people (whether sociocultural understandings consider them phenotypically different or similar), researchers silence details about smaller regional admixture processes. From an epistemological point of view, we can think about these choices as methodological trade-offs. As old and new models and datasets are evaluated, one aspect we want to emphasize is the current need to carefully address the spatial and temporal dimensions that articulate "admixture" as a genomic modeling enterprise.

In the next section, we focus on the peopling of the Americas to illustrate how admixture events are being modeled through time and space in a migration process that is estimated to have started 20,000 to 15,000 BP, depending on what set of archaeological, genomic, or paleontological findings are used.[34] We focus on the human population events that have come to stand as significant to narrate and set in motion old and new models and research workflows about past migrations into North, Central, and South America (see table 3.1). We examine the mechanisms that make such events stable and the workflows that can challenge them. Certainly, our strategy overlaps, at a smaller scale, with Michel Foucault's critique of knowledge (e.g., historical) and sociocultural power. At the end of the 1970s, Foucault set in motion the concept of "eventalization" to capture a method to disrupt the self-evidence of historical constants (through the construction of events, universalities) by pointing out and visibilizing singularities.[35] In our analysis, we are tracking down specific re-situations of historical demographic events and the emergence of new ones as admixture mapping and paleogenomics practitioners explore new models and yield datasets and findings. What we call *eventualization*[36] in this chapter is the current state of innovation, validation, and contestation of models and evidentiary datasets about the movement and transformation of human populations in the not so new "New World." These are insights about events and findings that have not yet turned into "facts,"[37] yet they are well known among certain specialized and general audiences.

THE PEOPLING OF THE AMERICAS:
MODELING ADMIXTURE(S)

In this section, we use a few published research outcomes that have challenged paleontological, archaeological, anthropological, and historical datasets by analyzing both contemporary and aDNA samples to characterize the peopling of the Americas from a genomic perspective.[38] This innovative approach is increasingly capturing the attention of funding agencies and of mass media to a point where the potential of genomics to provide answers is positioned higher than the ones held by the disciplines mentioned above.[39] Such hype not only allows the potential sensationalization[40] of genomic findings but also adds a great deal of complexity to the ways in which new genomic events (a sequenced specimen, a proposed migratory route, a new genomic ancestry) are supposed to be contextualized with larger or smaller existing events, regardless of their disciplinary origin (see table 3.1). This is the eventualization process that we also want to highlight, as HPGA modeling emphasizes or relativizes temporal and geographic scales as scientists infer from datasets.

Admixture through Contemporary DNA

In mid-July 2012, *Science* featured a news article by Anna Gibbons summarizing recent genomic findings about the peopling of the New World published online a few weeks before in a *Nature* article titled "Reconstructing Native American Population History."[41] Gibbons, a seasoned scientific journalist, emphasized that the 66 authors co-led by David Reich (at Harvard University) and Andrés Ruiz-Linares (then at University College London) brought to the spotlight a debate between

TABLE 3.1 Large-scale migration events for modeling the peopling of the Americas

Event	Temporal dimensions / dates	Geographic dimensions / locations	Population names / Demonyms	Presumed state of genomic diversity
f	After 1502	The New World (today's North, Central, and South America)	First peoples, Native Americans, Indigenous people, or Amerindians; Europeans; Africans; racial categories; national categories; ethnic categories; minority categories; etc.	Admixed**
e	1502	Forced migration from Africa to the New World	Africans	Isolated*
d	1492	Migration from the Old World to the New World	Europeans	Isolated*
c	After 15,000 BP	North America, south- and eastbound	First peoples, Native Americans, Indigenous people, or Amerindians	Isolated*
b	15,000 BP	Beringia	First peoples	Isolated*
a	Before 16,000 BP	Siberia / East Asia	Siberians / Asians	Isolated*

* This state could change to admixed depending on researchers' temporal and geographic scale (e.g., global, continental, regional, local).

** However, exceptions take place if researchers are sampling *assigned* and/or *self-identified* minorities (or isolated groups).

HPGA geneticists and archaeologists, biological anthropologists, and linguists over whether the first inhabitants of the Americas arrived in one wave of migration or in more.[42] Some HPGA researchers at the turn of the twenty-first century had leaned toward a model featuring one large-scale migratory wave—or a single founding population—since dataset comparisons were read then as showing that contemporary Native Americans across the continent were genetically similar.[43] This model countered, for example, a three-wave model proposed by interdisciplinary research teams twenty-six years earlier.[44] Reich, Ruiz-Linares, and their colleagues merged several datasets into a composite one with DNA samples from 493 Native Americans representing 52 populations, and from 245 individuals representing 17 populations in Siberia (for a total of 738 samples). The analysis of 364,470 single nucleotide polymorphisms in each of the 738 genomes allowed the researchers to argue that there were at least *three* migratory waves (or streams, as the authors call them) represented by three ancestral populations. The earliest one followed a coastal southward migratory pattern, and the latter two followed eastward directions after crossing a now extinct land bridge known as Beringia. The authors of the 2012 study framed their general findings as backing up the three migration waves suggested by the original model using interdisciplinary datasets,[45] rather than the one-wave model proposed using mitochondrial DNA and Y chromosome DNA.

We now briefly turn our attention to the modeling and curating adjustments that were required in order to make Reich and Ruiz-Linares's dataset work with an admixture model. The authors had to parse all samples to identify potential segments of recent European and/or African admixture based on historical timelines (see table 3.1) and mathematically "mask" data—exclude some alleles and segments—for all of the samples that were subjected to analysis.[46] In other words, in this particular research study and its main research question (Who were the first Americans?), the basic admixture model that starts with two isolated populations was turned into a model with three supposedly isolated populations (i.e., First Peoples, Europeans, and Africans at the turn of the sixteenth century) and potentially not one but several new generations of admixed populations based on chromosomal differences (segments) inherited from these three ancestral populations, or "deep lineages," as the authors sometimes called them. Indeed, the study carried out by such a large network of researchers at the time could have been considered "the most comprehensive survey of genetic diversity in Native Americans so far,"[47] but the findings also allowed the researchers to infer "back-migration" events for some of the populations that embodied each of the three proposed migratory streams. We want to emphasize that this is also an interesting contribution to the theoretical modeling character of admixture itself, one that highlights that its descriptive/evidentiary prowess diminishes when forced to map out a higher resolution, beyond general migratory events at continental scales and in search of regional details.[48] Yet this is far from being an end for the model.

On the contrary, it is a threshold for the further development and exploration of algorithmic tools (e.g., software), higher genomic mapping resolutions, the search for new specimens and samples, new data compilations, and the recalibration of old and new diversity or AIM panels.

In hindsight, this pattern of innovation in HPGA during the 2010s is what has promoted the increasing extraction and exploration of aDNA from biological-archaeological specimens collected in past and current research projects across the Americas.[49]

Admixture through Ancient DNA

By 2016, new technical breakthroughs and research outcomes using contemporary Native American and aDNA samples granted David Reich and Pontus Skoglund an opportunity to update the overview of the peopling of the Americas offered by Reich et al. four years earlier. Based on different aDNA specimens and further analysis of other contemporary samples from Native Americans across the region,[50] Skoglund and Reich argued:

> It is now clear that so many founder events and fluctuations in population size have occurred before, during, and after the peopling of the Americas that the evidence from one position in the genome—mitochondrial DNA, the Y chromosome, or any other location—is too subject to random changes in frequency (genetic drift) to provide a complete picture by itself. Only by taking the independent testimony of many locations in the genome simultaneously can we obtain a high-resolution picture of the deep past. The remainder of this article focuses on insights from whole genome studies of Native American population history. While these studies are still in their early days, they have already upended our understanding of key events. Application of ancient DNA technology promises further insights in years to come.[51]

Both researchers highlighted that the value of aDNA lies in offering a more complex picture of past and contemporary genetic structures by filling geographic gaps and complicating currently debated inferences about streams of ancestry/migration.[52]

Indeed, Skoglund and Reich argued that one of the most novel insights coming from the whole-genome approach listed in the 2016 update was the presence of a statistical signal linking contemporary Native American groups in the Amazon and Australo-Melanesians and Andaman islanders. To interpret that link, researchers suggested the theoretical existence of an ancient lineage, *Population Y*—short for the Tupi word *Ypyuéra*, the closest equivalent for the concept of "ancestor."[53] In this inferring scenario, Population Y also represents a wave of migration that contributed to early Australasians and First Americans but not necessarily through a north-to-south model of migration across the Americas. Since none of the aDNA samples available for the Americas at the time (2016) showed this type of ancestry, the authors hypothesized that a population with a more Australasian-related ancestry may have been present "in the broad geographic area in order to contribute to the *founders* of Native American founders."[54] In other words, a three-layered source of ancestry. They also gestured at a possible alternative scenario: "pulses" of migration[55] through Beringia following a coastal migratory route and then through an ice-free corridor that allowed eastward exploration.

Conceptually, what all the new scattered genomic findings reported by Skoglund and Reich do for admixture modeling is signal that the premise of isolation for ancestral populations is very relative, particularly when aiming to understand populations and events that took place before the individuals represented by the aDNA inhabited the geographic area under analysis (e.g., Beringia first, then the Americas; or all the previous admixture population events prior to event *a* in table 3.1). Similarly, new smaller models need to be integrated within the larger admixture model with a view to start interpreting similarities and differences between patterns of genomic ancestry between individuals representing specific ancient or contemporary populations. Some of these models embed migratory patterns such as "pulses," "back-migrations," and "ghost populations,"[56] as represented in the case of Population Y.[57] Empirically, on the other hand, the emerging of newer aDNA datasets won't necessarily provide definite answers to larger and smaller genomic questions but most likely will challenge older, current, and newer inferences and the models behind them. Such a process may only be enhanced when archaeologists, biological anthropologists, and/or linguists aim to contextualize such datasets vis-à-vis their own, a painstaking process that itself justifies the production of multiple literature reviews and overviews around the main research question. This contrasts with what life scientists think new findings about the research question should generate.[58]

Both Skoglund and Reich generously stated for the audiences of the journal *Current Opinion in Genetics and Development* that "a true understanding of the population history of any group or region cannot be achieved through genomic studies alone, but requires a synthesis of insights from genomics with information from anthropology, linguistics, archaeology and sociology."[59] However, the conscious or unconscious use of the word *information* in the previous sentence rather than *data* or *insights* to describe research outcomes from areas other than genomics, or from multidisciplinary, interdisciplinary projects, points to larger disciplinary tensions that have been emerging as aDNA—a substance—is turned by life scientists into a research area in its own right (i.e., paleogenomics) and extending beyond the scope of *Homo sapiens*. The core of the tension centers around the idea that, for some life scientists in HPGA, genomic datasets are more informative and valuable than paleontological, archaeological, and historical datasets.[60] This alleged authority to "rewrite history" influences not only how concepts like genomic admixture or genomic ancestry are instrumentalized in research workflows that include or exclude assumptions, models, and findings from other disciplines (e.g., anthropology, archaeology, linguistics, paleontology), but also how findings produced with those concepts are set in motion to travel among several audiences. Most of the time, such impacts take the form of oversimplifications, essentializations, and misunderstandings that can point out key epistemological challenges that intersect the workflows of archaeologists, biological anthropologists, geneticists, and paleontologists alike.

A concrete example of such common challenges for HPGA researchers involves the production and use of metadata—again, broadly understood for the purpose of our analysis as data about data—to characterize and represent the individuals behind contemporary and aDNA samples and the larger groups they are supposed to represent. In table 3.1 we have listed a few key large-scale migration events that several generations of researchers interested in the peopling of the Americas have produced and used as a temporal and geographical grid for the exploration of new models,[61] the contextualization of new findings, and the production of inferences. Each event, from the oldest (*a*) to the most recent (*f*), is associated with modeling (presumed) state(s) of genomic diversity that in turn enabled the identification of larger patterns of genomic transformation and usually delivered results as ancestry percentages. Each association requires assumptions from other disciplines that allow HPGA researchers to circumscribe DNA samples as representative of specific groups of people and no others. Likewise, it requires assumptions to circumscribe segments within such DNA sequences as representative of past ancestries to other groups of people through time and space. The following considerations are only a few structural conceptual challenges HPGA researchers should engage when carrying out research workflows and interrogating newer findings.

If we consider migration event (*b*), for example, we can ask when it makes sense to stop calling migrants exploring Beringia "Siberians" and start calling them "First Peoples" or "Native Americans." Archaeologists and paleontologists have offered evidence of how to trace and differentiate these kinds of transformations through time, but such datasets do not necessarily overlap with the patterns of difference offered by genomic analyses, either for aDNA specimens or for contemporary genomes.

If we consider migration events (*d*) and (*e*), researchers focus on complex genomic admixture processes that started taking place in the sixteenth century and that in the present can be reduced to the characterization that some contemporary populations in the Americas are admixed (e.g., "Latin Americans," as viewed from North America). What we want to emphasize here is that modeling at such large temporal and geographic scales pushes to the background the admixture stories behind standard categories such as "Native Americans," "Europeans," and "Africans." In other words, it is well supported through archaeological, bioanthropological, and historical records that such groups were not genomically homogeneous at the moment of their encounter. On the contrary, each of them represented complex states of genomic diversity or multiple genomic ancestries.

If we consider the migration events contained in (*f*) as a point of departure to model and reconstruct ancestral populations, the admixed characterization of certain groups of people across the Americas can quickly become less tangible. For this, HPGA researchers only need to increase the geno-graphic resolution and pay attention to the historical regional migrations and breeding processes that substantiate the emergence of countries' populations represented through individual samples in regional studies and composite datasets.[62] The genomic diversity reported with larger datasets for each country contest the assumption that "Latin Americans" show a homogeneous admixture genomic pattern.[63]

Similarly, this general pattern gets challenged when researchers ponder the fact that there are populations in each of these countries that consider themselves *direct* descendants, without admixture, of one of the three ancestral/continental populations. Should these self-reported distinct populations be included as separate populations or as "versions" of the continental ancestries looked for in "admixed" individuals analyzed by HPGA researchers working in the region? Would such a maneuver make sense from a modeling/biostatistical point of view? As we have seen, for Reich and some of his colleagues, this approach could work if the right algorithmic tools are deployed to exclude or mask certain genomic ancestries for recent admixture events, as they did for Native American samples.[64]

TRACKING THE RE-SITUATION OF ADMIXTURE(S)

In 2018 Reich published a book written for general and specialized audiences that aimed to answer both questions he used as a title: *Who We Are and How We Got Here: Ancient DNA and the New Science of the Human Past.*[65] The subtitle already elevated aDNA from a biological material with evidentiary power into *the science* that could answer the previous questions. Reich's justification for such a claim was the "ancient DNA revolution" that he and his lab members had helped to forge in the previous years by making aDNA "industrial."[66] By that he meant the streamlining of extraction, sorting, and sequencing fragments of aDNA that allowed the comparison of individuals at a pace not experienced before and that quickly surpassed previous article production and publication. Although Reich encouraged his readers to avoid processing the overview of the revolution and the answers to the research questions as definite (based on some findings and inferences), the tone, as he moves in chapters through continental areas, is dismissive, for example, of what archaeologists and biological anthropologists have contributed and about what they might contribute in the future. This we read as counterproductive at times, considering that it is in an interdisciplinary counterpointing of models, workflows, datasets, and findings where the discussion of concepts such as genomic admixture can yield more robust outcomes for all researchers invested in HPGA. A concrete example of this is the way in which archaeological datasets can relativize genomic ones (i.e., inferred ancient ancestries from a few or single specimens) by showing that these do not necessarily represent well-circumscribed groups—in sociocultural terms—through space and time.

The overall arguments in the book were, by design, meant to be controversial at many levels. One expected controversy exploded on March 23, 2018, when the *New York Times* Sunday Review published excerpts of the book, mostly from chapter 11, "The Genomics of Race and Identity."[67] The Sunday Review piece was titled "How Genetics Is Changing Our Understanding of Race." With some adjustments to make it work as a stand-alone piece, Reich boldly argued that, as a geneticist working on contemporary and ancient DNA, he knew "that it is simply no longer possible to ignore average genetic differences among races."[68] He did state that race was *also* a social construction, but argued that this characterization had been turned by well-meaning researchers into an orthodoxy rooted in "outdated" genetic studies by figures like Richard C. Lewontin (1929–2021). Lewontin had found in 1972 that most genetic variation was linked to differences between individuals and only a very small fraction was linked to differences among racial (continental) groups represented in his study.[69] Reich argued further that, because of the fast-paced turnout of data/evidence he and his colleagues were producing, human civilization should be prepared for more *findings* showing that differences among populations do exist.[70] Needless to say, the speed with which the book and digital versions of the abbreviated piece traveled was matched only by equally fast reactions, debates, and counterarguments,[71] and promoted more attention to the book and further interrogation of its structure and evidentiary foundation. Producing a detailed map of the impact of Reich's arguments on race could yield significant insights on multiple topics but would derail our main purpose in this chapter. However, we must say that Reich's positioning on the concept of race and racism in the book and in the Sunday Review stood in high contrast with the previous scientific reports on admixture and the peopling of the Americas, in which both were absent.[72]

In the previous scientific reports that we used to build and articulate some of our reflections, Reich and colleagues kept a somehow straightforward and overall consistent use of the concept of genomic admixture and a cautious take on the yields of admixture mapping and loci ancestry assignment to answer the questions of who the humans who made it to the Americas were and how they did so. Yet Reich's individual voice in the book overemphasized the current and future descriptive power of aDNA and the predictive power of genomics for medicine and/or highly contentious research areas linking biology to cognition and/or behavior and socioeconomic status:[73] "With the help of [DNA sequencing technology], we are learning that while race may be a social construct, differences in genetic ancestry that happen to correlate to many of today's racial constructs are real."[74] Furthermore, as he weaves an argument for genomic differences between contemporary populations, we read implicit conceptual maneuvers that render opaque the concept of *admixture*, for both ancient and recent samples.

This is an argumentative process in which patterns of ancient genomic ancestry are read as waves or streams of migrations not only to *interpret* genomic structure between specimens but also to substantiate theoretical populations into clearly circumscribed ones—culturally first and now racially. We say theoretical populations because admixture patterns can go beyond the genome of the specimen under consideration and be used to make inferences about genomic ancestry that are not yet substantiated with material culture (e.g., Population Y, or "ghost populations" in the context of the peopling of the Americas). From another angle, Reich's arguments about biological differences among current populations can be read as elevating *findings* to *facts* in order for genomics to have larger stakes in terms of predictive power. Arguing that aDNA studies show an "exquisite accuracy"[75] undoes the cautious approach followed by Reich and his coauthors in the research papers mentioned above, where they explored, inferred, and contextualized genomic ancestry data with other types of datasets. When concepts such as ancient "Native American ancestry" are re-situated and used to substantiate the idea of a contemporary homogeneous population—in economic, sociocultural, and now racial terms across the Americas—we witness how the overall finding of *continual* ancient admixtures is rendered, at best, opaque. (See Lisa Ikemoto's chapter for an analysis of case studies in which the idea of "pure" races is revealed as a fiction.) In this context, the impossibility of tracking other scientific objects and assumptions, dropped in the process of re-situation, opened doors for multiple audiences to doubt and question the robustness of the overall workflow of HPGA and HPGB as represented in Reich's individual work.[76]

Reich's own re-situation of some of the research findings he produced in collaboration with colleagues—in multiple institutions and countries—to back up his individual perspectives about the biological differences between populations shows us several important things. In terms of the concept of genomic admixture, the content and fate of the Sunday Review piece show us that, although the core goal of the model is to reconstruct sources of genomic ancestry regardless of what a past ancestral and admixed population looked like—the phenotypes of its members[77]—genomic admixture as a concept (and as a process) can be easily oversimplified and misinterpreted as synonymous with racial mixing. What this facilitates is substantiating HPGA findings to corroborate lay stereotypes about "racial populations" and correlations with disease prevalence and biomedical risk, such as those discussed in Tina Rulli's chapter and in the conclusion to this volume. The oversimplification of genomic admixture and the misinterpretations that ensue don't take place only when life scientists themselves re-situate components of their workflows. HPGA findings are re-situated by multiple audiences. A relevant example of this type of re-situation was set in motion with the publication of British journalist Nicholas Wade's book *A Troublesome Inheritance: Genes, Race and Human History* by Penguin Press.[78] In this book, the long-term contributor to the science section of the *New York Times* re-situated multiple HPGA authors, findings, and debates and elevated some of them into reiterations of racial stereotypes about behavior.[79] We point out Wade's case here because, interestingly enough,

Reich uses Wade's re-situation to illustrate "irresponsible" and "racist" stereotypes that have no genetic evidence backing them up.[80] Reich was even one of the 138 life scientists who drafted and signed an open letter published in the *New York Times Book Review*, some of whom had been cited by Wade.[81] At stake here is the need to compare in the future how both of these types of re-situation (by practitioners and by journalists) end up being re-situated in turn by other audiences and what is lost and gained, silenced, and enhanced in the process as they travel through society.[82]

CONCLUSION: ON HOW NOT TO OVERSIMPLIFY GENOMIC ADMIXTURE

Reich's own re-situation of some of the workflows he co-produced with multiple researchers allows us to gesture at the epistemological frailty of re-situating concepts, models, and findings in HPGA. In order to make an argument about the biological basis of some of the global racial assumptions behind populations, and to frame genomic descriptive power into predictive power for biomedical purposes, Reich set findings to travel without other companion objects (e.g., models and assumptions) to a point where they lost robustness. HPGA knowledge requires that the objects in HPGA workflows travel in bundles in order to maintain clarity about the exploratory character of the research area. When that doesn't happen, objects such as findings sit far from the confirmatory position they are expected to have in larger debates (from the peopling of the Americas to the biological and genetic basis of race). Our aim is to blame not the re-situations of genomic knowledge in general but, as we have seen, the *conditions* of it. The settings in which knowledge is set to travel could be improved by taking humble steps that could help audiences gain clarity about how HPGA workflows have been designed and therefore insights into what the extent of inferences could be for objects like models, datasets, and findings. In the specific cases of HPGA admixture mapping studies, one such step could be for researchers to make explicit in their scientific articles and publications how the concept of genomic admixture is being deployed conceptually and practically. This should also include the process of *ancestry assignment*[83] for both ancient and contemporary samples and the larger production of inferences about the populations such samples are supposed to represent.

Our proposal is inspired by more general calls inviting life scientists to clearly state in their research outcomes how they are understanding and using concepts such as race and ethnicity in HPGB and HPGF.[84] These calls have been motivated overall to address biases in how the two concepts—turned in the twentieth century into governmental categories at a global scale—have been granted at the same time too much descriptive and attributive power to assess health risk. Descriptive power in the sense that both categories have been used to identify economic, sociocultural, and political conditions that qualify health outcomes and health infrastructure. Attributive power, on the other hand, in the sense that race and ethnicity have been used to point out causal mechanisms in disease prevalence. Furthermore, both instances have a looping effect: descriptive uses can easily reinforce attributive explanations that inappropriately inform clinical applications. The social spaces that this transformation aimed to impact were multiple, including schools where new generations of health practitioners are being trained; public and private organizations overseeing provision of health services and/or grant funding; and publishers in control of scientific journals. Although the outcomes of all these efforts in the last two decades varies greatly from country to country due to local challenges that will require scientists to adopt similar models to enforce research and clinical guidelines, it is fair to point out that editors and editorial boards of scientific and biomedical journals have shown willingness and have taken concrete actions toward requiring authors to clarify and justify their use of racial and ethnic categories.[85] In retrospect, perhaps the most positive outcome of these efforts has been to keep the discussion alive so that editorial boards continue revising and updating their guidelines on how to report both categories.[86]

Going back to HPGA and the concept of admixture, we would like to see specialized journals require authors to elaborate on how the concept is conceived and incorporated into their workflows so as to ensure transparency in a process that very easily can be rendered opaque, as authors aim to elevate the exploration of migratory and genomic ancestry models into findings. We anticipate that clarifications about the concept of genomic admixture and the populational processes it seeks to track may not by itself suffice to ensure transparency. In most cases, clarification about admixture would also need clarification and contextualization about other theoretical concepts that, more often than not, pass unaddressed in both specialized and public opinion. This is the case for *population*. Population(s), understood at a theoretical level required for modeling genetic admixture to signify a group of individuals in which random mating is possible, at times also signifies a group of individuals that share sociocultural or biological characteristics. Such nuances facilitate the oversimplification of concepts and genomic processes, particularly when weaving datasets and findings from other disciplines' workflows into one's own work. As we have previously pointed out, oversimplifications open doors for misunderstandings and reifications (e.g., about *ancestry* in general, about genomic ancestry and sociocultural identity at both collective and individual levels). However, scientific and biomedical journals have predesignated sections for all the different types of articles they publish (e.g., methods, discussion, and online supplemental materials), where HPGA researchers can make explicit how they understand and use concepts like genomic admixture and population and the way they produce findings. Unfortunately, these sections (other than "discussion") are not likely to be read by readers and *re-situators* in audiences outside the small circle of specialists who conduct this kind of research.

We are not naïve about expecting that the inclusion of detailed reflections about scientists' concepts will prevent misunderstandings in how multiple actors could re-situate findings or any other isolated aspect of the workflow. Or, for that matter, how researchers decide to re-situate aspects of their own research outcomes (e.g., Reich). However, having clarifications about how genomic admixture was used and about what cannot be inferred from it affords the possibility of revisiting aspects of the workflow that could prevent other scientists, critics, and consumers of HPGA knowledge from misinterpreting findings and narratives.

Likewise, if misunderstandings and oversimplifications have taken place, such a record in future publications will allow third parties to track down how and where they happened. In other words, there will be an opportunity to partially reconstruct the process of re-situation and find the conceptual and analytical seams that were turned into oversimplifications and potential misunderstandings. Several life scientists have considered and embraced this as an approach that enhances both the robustness and the ethics of their research activities, particularly when working among vulnerable communities.[87]

Bringing clarity to concepts like genomic admixture (or population, for that matter) keeps expanding discussion spaces already opened by the questioning of race and ethnicity as meaningful categories in research areas such as HPGA, HPGB, and HPGF. This is a small but much-needed step, since the boundaries between genomic ancestry and biomedical data are being blurred by some life scientists and by private DTC companies to a point where public opinion reads ancestry and genomic ancestry mostly in terms of disease risk. For the particular research question about the peopling of the Americas, these efforts will prevent misinterpreting the current theoretical scaffold around ancient migration events (see table 3.1, events *a* to *e*) in order to substantiate individual and collective identity through the lenses of race (table 3.1, event *f*) or even ethnicity when used in theory as a politically correct synonym for race.

NOTES

Carlos Andrés Barragán's and James Griesemer's work on this topic is supported by the US National Science Foundation (NSF) under Grant No. SES-1849307–0. Any opinions, findings, and conclusions or recommendations expressed in this material are those of the author(s) and do not necessarily reflect the views of the NSF.

1. Dorothy Nelkin and M. Susan Lindee, *The DNA Mystique: The Gene as a Cultural Icon* (Ann Arbor: University of Michigan Press, 1995).

2. Kathy Hudson, Gail Javitt, Wylie Burke, and Peter Byers, "ASHG Social Issues Committee: ASHG Statement on Direct-to-Consumer Genetic Testing in the United States," *American Journal of Human Genetics* 81, no. 3 (2007): 635–37; Henry T. Greeley, "The Future of DTC Genomics and the Law," *Journal of Law, Medicine and Ethics* 48, no. 1 (2020): 151–60.

3. See Mark Fedyk's chapter in this volume for a complementary perspective on this topic.

4. See Keith Wailoo, Alondra Nelson, and Catherine Lee, eds., *Genetics and the Unsettled Past: The Collision of DNA, Race, and History* (New Brunswick, NJ: Rutgers University Press, 2012).

5. James R. Griesemer, "A Data Journey through Dataset-Centric Population Genomics," in *Data Journeys in the Sciences*, ed. Sabina Leonelli and Niccolò Tempini (Cham, Switzerland: Springer, 2020),

146–67; James R. Griesemer and Carlos Andrés Barragán, "Re-situations of Scientific Knowledge: A Case Study of a Skirmish over Clusters vs. Clines in Human Population Genomics," *History and Philosophy of the Life Sciences* 44 (2022), article 16.

6. Spencer Wells, *The Journey of Man: A Genetic Odyssey* (London: Allen Lane, 2002); *Journey of Man*, directed by Clive Maltby (Alexandria, VA: PBS Home Video and National Geographic Society, 2003); Spencer Wells, *Deep Ancestry: Inside the Genographic Project* (Washington, DC: National Geographic Society, 2006).

7. E.g., Jennifer Reardon, *Race to the Finish: Identity and Governance in an Age of Genomics* (Princeton, NJ: Princeton University Press, 2005); Catherine Bliss, "Mapping Race through Admixture," *International Journal of Technology, Knowledge and Society* 4, no. 4 (2008): 79–83; Bliss, *Race Decoded: The Genomic Fight for Justice* (Stanford, CA: Stanford University Press, 2012); Duana Fullwiley, "The Biologistical Construction of Race: 'Admixture' Technology and the New Genetic Medicine," *Social Studies of Science* 38, no. 5 (2008): 695–735; Sandra Soo-Jin Lee, Deborah A. Bolnick, Troy Duster, Pilar Ossorio, and Kim TallBear, "The Illusive Gold Standard in Genetic Ancestry Testing," *Science* 325, no. 5936 (2009): 38–39; Sahra Gibbon, Ricardo Ventura Santos, and Mónica Sans, eds., *Racial Identities, Genetic Ancestry and Health in South America* (New York: Palgrave Macmillan, 2012); Catherine Nash, *Genetic Geographies: The Trouble with Ancestry* (Minneapolis: University of Minnesota Press, 2015); Kim TallBear, *Native American DNA: Tribal Belonging and the False Promise of Genetic Science* (Minneapolis: University of Minnesota Press, 2013).

8. E.g., Deborah A. Bolnick, "Individual Ancestry Inference and the Reification of Race as a Biological Phenomenon," in *Revisiting Race in a Genomic Age*, ed. Barbara A. Koenig, Sandra Soo-Jin Lee, and Sarah S. Richardson (New Brunswick: Rutgers University Press, 2008), 70–85; Graham Coop, Michael B. Eisen, Rasmus Nielsen, Molly Przeworski, Noah Rosenberg, et al., letter to the editor, *New York Times*, Book Review, August 18, 2014, https://nytimes.com/2014/08/10/books/review/letters-a-troublesome-inheritance.html; Marcus W. Feldman and Richard C. Lewontin, "Race, Ancestry, and Medicine," in *Revisiting Race in a Genomic Age*, ed. Barbara A. Koenig, Sandra Soo-Jin Lee, and Sarah S. Richardson (New Brunswick, NJ: Rutgers University Press, 2008), 89–101; Ian Mathieson and Aylwyn Scally, "What Is Ancestry?," *PLoS Genetics* 16, no. 3 (2020): e1008624.

9. E.g., Deborah A. Bolnick, Duana Fullwiley, Troy Duster, Richard S. Cooper, Joan H. Fujimura, Jonathan Kahn, Jay S. Kaufman, Jonathan Marks, Ann Morning, Alondra Nelson, Pilar Ossorio, Jenny Reardon, Susan M. Reverby, and Kimberly TallBear, "The Science and Business of Genetic Ancestry Testing," *Science* 318, no. 5849 (2007): 399–400; Lee et al., "The Illusive Gold Standard"; Gibbon, Santos, and Sans, *Racial Identities*.

10. E.g., J. Claiborne Stephens, David Briscoe, and Stephen J. O'Brien, "Mapping by Admixture Linkage Disequilibrium in Human Populations: Limits and Guidelines," *American Journal of Human Genetics* 55, no. 4 (1994): 809–24; Paul M. McKeigue, "Prospects for Admixture Mapping of Complex Traits," *American Journal of Human Genetics* 76, no. 1 (2005): 1–7; Michael W. Smith and Stephen J. O'Brien, "Mapping by Admixture Linkage Disequilibrium: Advances, Limitations and Guidelines," *Nature Reviews Genetics* 6, no. 8 (2005): 623–32; Cheryl A. Winkler, George W. Nelson, and Michael W. Smith, "Admixture Mapping Comes of Age," *Annual Review of Genomics and Human Genetics* 11 (2010): 65–89; Line Skotte, Thorfinn Sand Korneliussen, and Anders Albrechtsen, "Estimating Individual Admixture Proportions from Next Generation Sequencing Data," *Genetics* 195, no. 3 (2013): 693–702; Katherine L. Korunes and Amy Goldberg, "Human Genetic Admixture," *PLOS Genetics* 17, no. 3 (2021): e1009374.

11. The topic has been formalized for a general audience—for example, in the *New York Times* with a popular column titled "Origins" under the direction of Carl Zimmer, a journalist and an author who specializes in evolution and heredity. See the scope of the column at https://www.nytimes.com/column/origins. He is also the author of *She Has Her Mother's Laugh: The Powers, Perversions, and Potential of Heredity* (New York: Dutton, 2018).

12. E.g., Bolnick et al., "Science and Business of Genetic Ancestry Testing"; Bolnick, "Individual Ancestry Inference"; Lee et al., "The Illusive Gold Standard"; Jenny Reardon and Kim TallBear, "'Your

DNA Is Our History': Genomics, Anthropology, and the Construction of Whiteness as Property," *Current Anthropology* 53, Supplement 5 (2012): S233–45.

13. E.g., Mónica Sans, "Admixture Studies in Latin America: From the 20th to the 21st Century," *Human Biology* 72, no. 1 (2000): 155–77; Indrani Halder and Mark D. Shriver, "Measuring and Using Admixture to Study the Genetics of Complex Diseases," *Human Genomics* 1, no. 1 (2003): 52–62; Francisco M. Salzano, "Interethnic Variability and Admixture in Latin America—Social Implications," *Revista de Biología Tropical* 52, no. 3 (2004): 405–15; Guilherme Suarez-Kurtz, ed., *Pharmacogenomics in Admixed Populations* (Austin: Landes Bioscience, 2007); Alkes Price, Nick Patterson, Fuli Yu, David R. Cox, Alicja Waliszewska, Gavin J. McDonald, Arti Tandon, Christine Schirmer, Julie Neubauer, Gabriel Bedoya, Constanza Duque, Alberto Villegas, Maria Cátira Bortolini, Francisco M. Salzano, Carla Gallo, Guido Mazzotti, Marcela Tello-Ruíz, Laura Riba, Carlos A. Aguilar-Salinas, Samuel Canizales-Quinteros, Marta Menjivar, William Klitz, Brian Henderson, Christopher A. Haiman, Cheryl Winkler, Teresa Tusie-Luna, Andrés Ruiz-Linares, and David Reich, "A Genome Wide Admixture Map for Latino Populations," *American Journal of Human Genetics* 80, no. 6 (2007): 1024–26; Winkler, Nelson, and Smith, "Admixture Mapping Comes of Age."

14. For social scientific and STS analyses of the impacts of DTC tests, see, for example, Barbara A. Koenig, Sandra Soo-Jin Lee, and Sarah S. Richardson, eds., *Revisiting Race in a Genomic Age* (New Brunswick, NJ: Rutgers University Press, 2008); Wailoo, Nelson, and Lee, *Genetics and the Unsettled Past*; Reardon and TallBear, "'Your DNA Is Our History.'"

15. We understand re-situation as a process of accommodating the direct or indirect transfer of objects of knowledge from one site/situation to (one or many) other sites/situations. See James R. Griesemer and Carlos Andrés Barragán, Standard Grant: A Case Study of How Re-Situation of Scientific Knowledge from Human Population Genomics Works, NSF grant SES-1849307, 2019–2024; Griesemer and Barragán, "Re-situations of Scientific Knowledge."

16. Following Peter Howlett and Mary S. Morgan, eds., *How Well Do Facts Travel? The Dissemination of Reliable Knowledge* (Cambridge: Cambridge University Press, 2011).

17. See Jayne O. Ifekwunigwe, ed., *'Mixed Race' Studies: A Reader* (New York: Routledge, 2004); Carlos Andrés Barragán, "Untangling Population Mixture? Genomic Admixture and the Idea of *Mestizos* in Latin America," *Gene Watch* 28, no. 2 (2015): 11–13, 20–21.

18. Mark Jobling, Edward Hollox, Toomas Kivisild, and Chris Tyler-Smith, *Human Evolutionary Genetics*, 2nd ed. (New York: Garland Science, [2004] 2014), 609.

19. Jobling et al., *Human Evolutionary Genetics*.

20. *Oxford English Dictionary*, accessed February 8, 2024, https://oed.com.

21. E.g., Richard G. Klein, *The Human Career: Human Biological and Cultural Origins* (Chicago: University of Chicago Press, [1989] 2009); Luigi Luca Cavalli-Sforza, Menozzi Piazza, and Alberto Piazza, *The History and Geography of Human Genes* (Princeton, NJ: Princeton University Press, 1994).

22. Some life scientists refer to *ancestral populations* also as *parental populations*. In this chapter, we decided to use the former term since it avoids overemphasizing kinship metaphors that can obscure the uses of genomic admixture as a model at different timescales.

23. Mathieson and Scally, "What Is Ancestry?"

24. An example of this is when HPGA models, methods, and findings have been re-situated by other life scientists to other areas such as HPGB and HPGF.

25. E.g., Luigi Luca Cavalli-Sforza, Allan C. Wilson, Charles R. Cantor, Robert M. Cook-Deegan, and Mary-Claire King, "Call for a Worldwide Survey of Human Genetic Diversity: A Vanishing Opportunity for the Human Genome Project," *Genomics* 11, no. 2 (1991): 490–91; see also Luigi Luca Cavalli-Sforza, "Human Genome Diversity: Where Is the Project Now?," in *Human DNA: Law and Policy*, ed. Bartha Maria Knoppers, Claude M. Laberge, and Marie Hirtle (Boston: Kluwer Law International, 1997), 219–27.

26. In a very general sense, *linkage analysis* is the mapping of genes (using pedigrees, for example) or traits on the basis of their tendency to be co-inherited with polymorphic loci. On the other hand, the

concept of *linkage disequilibrium* (LD) describes the nonrandom association between alleles in a given population because of the tendency to be co-inherited.

27. E.g., David C. Rife, "Populations of Hybrid Origin as Source Material for the Detection of Linkage," *American Journal of Human Genetics* 6, no. 1 (1954): 26–33; Ranajit Chakraborty and Kenneth M. Weiss, "Admixture as a Tool for Finding Linked Genes and Detecting That Difference from Allelic Association between Loci," *Proceedings of the National Academy of Sciences* 85, no. 23 (1988): 9119–23; Winkler, Nelson, and Smith, "Admixture Mapping Comes of Age."

28. Stephens et al., "Mapping by Admixture Linkage Disequilibrium"; Winkler, Nelson, and Smith, "Admixture Mapping Comes of Age."

29. E.g., Paul M. McKeigue, "Mapping Genes That Underlie Ethnic Differences in Disease Risk: Methods for Detecting Linkage in Admixed Populations, by Conditioning on Parental Admixture," *American Journal of Human Genetics* 63, no. 1 (1998): 241–51; Winkler, Nelson, and Smith, "Admixture Mapping Comes of Age."

30. Here it is important to distinguish between *ancestry assignment* and *admixture mapping*. The former, also referred to as *admixture proportion*, describes the process of estimating in a sample the proportion of the genome that is assigned to a particular ancestry. It requires the use of AIM and diversity panels, their doctoring, or the establishment of new ones.

31. E.g., Heather E. Collins-Schramm, Bill Chima, Takanobu Morii, Kimberly Wah, Yolanda Figueroa, Lindsey A. Criswell, Robert L. Hanson, William C. Knowler, Gabriel Silva, John W. Belmont, and Michael F. Seldin, "Mexican American Ancestry-Informative Markers: Examination of Population Structure and Marker Characteristics in European Americans, Mexican Americans, Amerindians and Asians," *Human Genetics* 114, no. 3 (2004): 263–71.

32. Carlos Andrés Barragán, "Molecular Vignettes of the Colombian Nation: The Place(s) of Race and Ethnicity in Networks of Biocapital," in *Racial Identities, Genetic Ancestry and Health in South America*, ed. Sahra Gibbon, Ricardo Ventura Santos, and Mónica Sans (New York: Palgrave Macmillan, 2012), 41–68.

33. See Susanne Hummel, *Ancient DNA Typing: Methods, Strategies and Applications* (Berlin: Springer-Verlag, 2003); Beth Shapiro, Axel Barlow, Peter D. Heintzman, Michael Hofreiter, Johanna L. A. Paijmans, and André E. R. Soares, eds., *Ancient DNA: Methods and Protocols*, 2nd ed. (New York: Human Press, [2012] 2019); Charlotte Lindqvist and Om P. Rajora, *Paleogenomics: Genome-Scale Analysis of Ancient DNA* (Cham, Switzerland: Springer, 2019).

34. This is a very active and therefore contested temporal model. New paleontological, archaeological, bioanthropological, and genomic findings are frequently emerging and framed as providing new evidence or resolving debates. For example, at the time of preparation of this manuscript, *Science* published an article reporting that the ancient human footprints found in what today is White Sands National Park in New Mexico were left as early as 23,000 years ago. Using radiocarbon analysis on conifer pollen seeds associated with the footprints, researchers confirmed that the footprints are evidence of earlier human entrance(s) to North America, before the well-documented migrations through the Beringia land bridge that occurred at the end of the Ice Age. Jeffrey S. Pigati, Kathleen B. Springer, Jeffrey S. Honke, David Wahl, Marie R. Champagne, Susan R. H. Zimmerman, Harrison J. Gray, Vincent L. Santucci, Daniel Odess, David Bustos, and Matthew R. Bennett, "Independent Age Estimates Resolve the Controversy of Ancient Human Footprints at White Sands," *Science* 382, no. 6666 (2023): 73–75. The workflow in this project was a response to skeptical reactions by critics to previous research findings that proposed a similar timeline but used radiocarbon analysis of aquatic plant seeds. Matthew R. Bennett, David Bustos, Jeffrey S. Pigati, Kathleen B. Springe, Thomas M. Urban, Vance T. Holliday, Sally C. Reynolds, Marcin Budka, Jeffrey S. Honke, Adam M. Hudson, Brendan Fenerty, Clare Connelly, Patrick J. Martinez, Vincent L. Santucci, and Daniel Odess, "Evidence of Humans in North America during the Last Glacial Maximum," *Science* 373, no. 6562 (2021): 1528–31.

35. Michel Foucault, "Questions of Method," in *Power: Essential Works of Foucault, 1954–1984*, vol. 3, ed. James D. Faubion (New York: The New Press, [1978] 2000), 223–38.

36. The difference in spelling is on purpose to signal the reader that our take on the concept is different from Foucault's.

37. See Howlett and Morgan, *How Well Do Facts Travel?*; Sabina Leonelli and Niccolò Tempini, eds., *Data Journeys in the Sciences* (Cham, Switzerland: Springer, 2020).

38. In general, contemporary DNA samples in the studies we dialogue with come from individuals representing Siberians, Native Americans and/or Indigenous peoples, and "admixed" populations across the Americas. Ancient DNA samples, on the other hand, come from several archaeological sites and specimen repositories located across these regions. For a theoretical and methodological view of aDNA research, see Hummel, *Ancient DNA Typing*; Shapiro et al., *Ancient DNA*; Lindqvist and Rajora, *Paleogenomics*.

39. See Michael Balter, "New Mystery for Native American Origins," *Science* 349, no. 6246 (2015): 354–55; Ewen Callaway, "Migration to Americas Traced," *Nature* 563, no. 7731 (2018): 303–4.

40. For a philosophy of biology take on sensationalization and aDNA, see Joyce C. Havstad, "Sensational Science, Archaic Hominin Genetics, and Amplified Inductive Risk," *Canadian Journal of Philosophy* 5, no. 3 (2022): 295–320.

41. Ann Gibbons, "Genes Suggest Three Groups Peopled the New World," *Science* 337, no. 6091 (2012): 144; David Reich, Nick Patterson, Desmond Campbell, Arti Tandon, Stéphane Mazieres, Nicolas Ray, Maria V. Parra, Winston Rojas, Constanza Duque, Natalia Mesa, et al., "Reconstructing Native American Population History," *Nature* 488, no. 7411 (2012): 370–75.

42. The findings of Reich et al. were also highlighted by other journalists in other scientific and news outlets; see, for example, Linda Geddes and Michael Marshall, "Once, Twice, Thrice into the Americas," *New Scientist* 215, no. 2873 (2012): 12; Robert Lee Holtz, "Early Americans Arrived in Three Waves," *Wall Street Journal*, July 12, 2012; Nicholas Wade, "Earliest Americans Arrived in Waves, DNA Study Finds," *New York Times*, July 12, 2012.

43. E.g., Erika Tamm, Toomas Kivisild, Maere Reidla, Mait Metspalu, David Glenn Smith, Connie J. Mulligan, Claudio M. Bravi, Olga Rickards, Cristina Martinez-Labarga, Elsa K. Khusnutdinova, et al., "Beringian Standstill and Spread of Native American Founders," *PLOS One* 2, no. 9 (2007): e829; Nelson J. R. Fagundes, Ricardo Kanitz, Roberta Eckert, Ana C. S. Valls, Mauricio R. Bogo, Francisco M. Salzano, David Glenn Smith, Wilson A. Silva Jr., Marco A. Zago, Andrea K. Ribeiro-Dos-Santos, Sidney E. B. Santos, Maria Luiza Petzl-Erler, and Sandro L. Bonatto, "Mitochondrial Population Genomics Supports a Single Pre-Clovis Origin with a Coastal Route for the Peopling of the Americas," *American Journal of Human Genetics* 82, no. 3 (2008): 583–92.

44. Joseph H. Greenberg, Christy G. Turner II, and Stephen L. Zegura, "The Settlement of the Americas: A Comparison of the Linguistic, Dental, and Genetic Evidence," *Current Anthropology* 27, no. 5 (1986): 477–97; see also Ann Gibbons, "Geneticists Trace the DNA Trail of the First Americans," *Science* 259, no. 5093 (1993): 312–13.

45. E.g., Greenberg, Turner, and Zegura, "Settlement of the Americas."

46. Reich et al., "Reconstructing Native American Population History," 374.

47. Reich et al., "Reconstructing Native American Population History," 373.

48. This type of development is hardly captured in news articles due to its abstract character. Yet tracking it turns out to be very useful for understanding how modeling assumptions evolve while following a workflow to reconstruct genomic admixture.

49. See David J. Meltzer, *First Peoples in a New World: Colonizing Ice Age America* (Berkeley: University of California Press, 2009); David Reich, *Who We Are and How We Got Here: Ancient DNA and the New Science of the Human Past* (New York: Pantheon Books, 2018); Jennifer Raff, *Origin: A Genetic Story of the Americas* (New York: Twelve Books, 2022).

50. Pontus Skoglund, Swapan Mallick, Maria Cátira Bortolini, Niru Chennagiri, Tábita Hünemeier, Maria Luiza Petzl-Erler, Francisco Mauro Salzano, Nick Patterson, and David Reich, "Genetic Evidence for Two Founding Populations of the Americas," *Nature* 525, no. 7567 (2015): 104–8.

51. The authors were also highlighting the value of pursuing whole genome sequencing since it produces multilocus rather than single-locus data. See Pontus Skoglund and David Reich, "A Genomic View of the Peopling of the Americas," *Current Opinion in Genetics and Development* 41 (2016): 27–35, 28.

52. This is a point that other life scientists also share, from a theoretical point of view, and have publicly declared: "Jennifer Raff, an anthropological geneticist at the University of Kansas in Lawrence, says that the emerging picture of the Americas is less a revision of the earlier models and more an elaboration. 'It's not that everything we know is getting overturned. We're just filling in details,' she says" (Raff, in Callaway, "Migration to Americas Traced," 304). See also Raff, *Origin*.

53. Skoglund et al., "Genetic Evidence for Two Founding Populations of the Americas," 106; Skoglund and Reich, "Genomic View of the Peopling of the Americas," 31–32.

54. Our emphasis. Skoglund and Reich, "Genomic View of the Peopling of the Americas," 31.

55. We interpret "pulses" in Skoglund and Reich's update article (2016) to be smaller and dispersed—in time and space—waves of migration.

56. Reich, *Who We Are*, 81–83, 96–97.

57. Skoglund et al., "Genetic Evidence for Two Founding Populations of the Americas"; Skoglund and Reich, "Genomic View of the Peopling of the Americas."

58. E.g., Raff, in Callaway, "Migration to Americas Traced," 304; Raff, *Origin*.

59. Skoglund and Reich, "Genomic View of the Peopling of the Americas," 33.

60. Reich argues in connection with the concept of "race" that "To understand why it is no longer an option for geneticists to lock arms with anthropologists and imply that any differences among human populations are so modest that they can be ignored, go no further than the 'genome bloggers.' [Their] political beliefs are fueled partly by the view that when it comes to discussion about biological differences across populations, the academics are not honoring the spirit of scientific truth-seeking. The genome bloggers take pleasure in pointing out contradictions between the politically correct messages academics often give about the indistinguishability of traits across populations and their papers showing that this is not the way the science is heading." Reich, *Who We Are*, 254–55.

61. This is far from being an exhaustive effort to group all the migratory events that can matter for HPGA researchers working in the region, and therefore it shouldn't be read as representing a settled analytical scaffold.

62. E.g., Reich et al., "Reconstructing Native American Population History"; see also Barragán, "Untangling Population Mixture?"

63. See, for example, the findings reported by two studies led by Andrés Ruiz-Linares back then at University College London: Sijia Wang, Nicolas Ray, Winston Rojas, Maria V. Parra, Gabriel Bedoya, Carla Gallo, Giovanni Poletti, et al., "Geographic Patterns of Genome Admixture in Latin American Mestizos," *PLOS Genetics* 4, no. 3 (2008): 1–9; Andrés Ruiz-Linares, Kaustubh Adhikari, Victor Acuña-Alonzo, Mirsha Quinto-Sanchez, Claudia Jaramillo, William Arias, Macarena Fuentes, et al., "Admixture in Latin America: Geographic Structure, Phenotypic Diversity and Self-Perception of Ancestry Based on 7,342 Individuals," *PLOS Genetics* 10, no. 9 (2014): e1004572. For an analysis of these two studies, see Carlos Andrés Barragán, "Lineages within Genomes: Situating Human Genetics Research and Contentious Bio-Identities in Northern South America" (PhD diss., University of California, Davis, 2016).

64. Reich et al., "Reconstructing Native American Population History."

65. Reich, *Who We Are*.

66. Reich, *Who We Are*, xix.

67. Reich, *Who We Are*, 247–73.

68. David Reich, "Race in the Age of Modern Genetics," *New York Times*, Sunday Review, March 23, 2018, https://nytimes.com/2018/03/23/opinion/sunday/genetics-race.html. Title of the online version: "How Genetics Is Changing Our Understanding of 'Race.'"

69. Reich referred to Lewontin's analysis of protein types in blood from individuals representing so-called racial groups—Africans, Asians (East), Asians (South), Australians, Eurasians (West),

Native Americans, and Oceanians—and found that 85 percent of the variation in protein types could be accounted for by variation within racial groups and only 15 percent among the groups. Richard C. Lewontin, "The Apportionment of Human Diversity," in *Evolutionary Biology*, ed. Theodosius Dobzhansky, Max K. Hecht, and William C. Steere (New York: Springer, 1972), vol. 6, 381–98. Also see Richard C. Lewontin, *Biology as Ideology: The Doctrine of DNA* (New York: Harper Perennial, [1991] 1992).

70. He elaborated more on this argument a week later in a short follow-up comment in the *New York Times* in which he addressed a few of the comments posted to the online version of his first article. See David Reich, "How to Talk about 'Race' and Genetics," *New York Times*, March 30, 2018, https://nytimes.com/2018/03/30/opinion/race-genetics.html.

71. For an example of how sociocultural anthropologists received Reich's arguments, see Jonathan Kahn et al., "How Not to Talk about Race and Genetics," *BuzzFeed News*, March 30, 2018, https://buzzfeednews.com/article/bfopinion/race-genetics-david-reich. Other public figures' reactions were compiled by the editor of the *New York Times* as "Race, Genetics and a Controversy," *New York Times*, April 2, 2018, https://nytimes.com/2018/04/02/opinion/genes-race.html.

72. See Reich et al., "Reconstructing Native American Population History"; Skoglund et al., "Genetic Evidence for Two Founding Populations of the Americas"; Skoglund and Reich, "Genomic View of the Peopling of the Americas."

73. See Emily Merchant's chapter in this volume for a thorough discussion of these topics.

74. Reich, "Race in the Age of Modern Genetics."

75. Reich, "Race in the Age of Modern Genetics."

76. See responses in Kahn et al., "How Not to Talk about Race and Genetics"; *New York Times*, "Race, Genetics and a Controversy."

77. It is important to clarify that in the context of "racial" takes to characterize current human populations, the phenotype is translated into an overgeneralization of some physical characteristics of individuals that are supposed to represent a given "race." In current biological theory, on the other hand, "phenotype" is understood as a set of observable characteristics in an individual resulting from interactions between the individual's genotype and the environment.

78. Nicholas Wade, *A Troublesome Inheritance: Genes, Race and Human History* (New York: Penguin, 2014).

79. Wade, *Troublesome Inheritance*.

80. Reich, "Race in the Age of Modern Genetics."

81. Coop et al., letter to the editor; see also Griesemer and Barragán, "Re-situations of Scientific Knowledge."

82. This is an endeavor we're working on, but expanding on it will overflow the scope and space for our analysis in this edited volume. However, we want to offer a glimpse of how the process of re-situation can turn into a rabbit hole. In response to Reich's Sunday Review article, Wade wrote a letter to the editor of the *New York Times* stating the following: "At last! A Harvard geneticist, David Reich, admits that there are genetic differences between human races, even though he puts the word race in quotation marks. Obvious as this may seem, American academics for decades have insisted that race is a social construct, and have vilified as a racist anyone who says otherwise. After covering the human genome project for this newspaper for many years, I wrote a book, *A Troublesome Inheritance*, which explained that there is indeed a biological basis to race, a fact that Mr. Reich has now echoed, though without acknowledgment. He even tries to portray my book as racist, which it is not. Acknowledging that race has a biological basis is a salutary advance. Opposition to racism should rest not on the lie that races don't exist but on principle, allowing science to proceed without hindrance. It is those who believe that free scientific inquiry will turn up something terrible who should check their consciences. The human genome's forceful message is one of unity: All races are but variations on a single theme." Wade, in the *New York Times*, "Race, Genetics and a Controversy." Wade seized the opportunity to use

Reich's re-situation of findings to legitimize arguments synthesized in his book (Wade, *Troublesome Inheritance*) and to dispute Reich's framing of Wade's book as racist. At stake in this brief skirmish is how each author values findings in HPGA and when it is objective to use such findings to substantiate differences between so-called "racial" populations.

83. See n. 30 above.

84. E.g., Simon M. Outram and George T. H. Ellison, "Anthropological Insights into the Use of Race/Ethnicity to Explore Genetic Determinants of Disparities in Health," *Journal of Biosocial Science* 38, no. 1 (2005): 83–102; Outram and Ellison, "Improving the Use of Race/Ethnicity in Genetic Research: A Survey of Instructions to Authors in Genetics Journals," *Science Education* 29, no. 3 (2006): 78–81.

85. See Outram and Ellison, "Improving the Use of Race/Ethnicity in Genetic Research"; Andrew Smart, Richard Tutton, Paul Martin, George T. H. Ellison, and Richard Ashcroft, "The Standardisation of Race and Ethnicity in Biomedical Science Editorials and UK Biobanks," *Social Studies of Science* 38, no. 3 (2007): 407–23; "Instructions for Authors: Reporting Demographic Information for Study Participants," *Journal of the American Medical Association*, accessed February 8, 2024, https://jamanetwork.com/journals/jama/pages/instructions-for-authors#SecReportingRace/Ethnicity; editorial, "Why *Nature* Is Updating Its Advice to Authors on Reporting Race or Ethnicity," *Nature* 616, no. 7956 (2023): 219.

86. E.g., editorial, "Why *Nature* Is Updating Its Advice to Authors."

87. E.g., Andrew Smart, Deborah A. Bolnick, and Richard Tutton, "Health and Genetic Ancestry Testing: Time to Bridge the Gap," *BMC Medical Genomics* 10 (2017), article 3; Mathieson and Scally, "What Is Ancestry?"

DNA and Reproduction

4

Selling Racial Purity
in Direct-to-Consumer Genetic Testing
and Fertility Markets

Lisa C. Ikemoto

Direct-to-consumer (DTC) genetic ancestry test companies and businesses that purvey human gametes provide carefully curated and bundled information that consumers can use to express and construct identity. These are genetic identity[1] markets. DTC genetic ancestry test companies offer reports that include verbal descriptions, charts, and quotients in exchange for a fee, personal information, and a spit sample. Sperm banks present a layered set of choices to intended parents that lead to selection of semen from a particular donor and all the traits attributed to the donor. Both markets use genetic ancestry in ways that code for race.

Both industries package identity in ways that prioritize the role of genetics and a genetic construction of race. In the twenty-first century, genetic race serves as a vehicle for a cluster of old ideas. This chapter elaborates on the updated versions of two old ideas. The first is racial purity, the idea that race remains intact, even after mixing. It is insoluble. In its distilled form, race can also be quantified, as seen in the chapter by Mark Fedyk. The analysis that follows traces the geneticized explanations for racial difference to the early nineteenth-century theory of polygenism, which will appear again in the chapter by Meaghan O'Keefe. The new polygenism does not necessarily claim that different racial populations have separate genetic origins, but it insists that genetic variations between racial populations are significant. It accommodates monogenism but accords greater significance to the racial ancestor than to the originating ancestor of humans.

What the new biomarkets offer is the purchase of fractionated racial identity, which is a vehicle for racial purity hidden behind a veneer of multiculturalism.

Sperm bank and genetic ancestry test company practices emphasize their ability to measure and quantify the racial components of identity, in ways that sum to 100 percent. Racial purity recalls old racisms that used science, albeit contested, to assert that the races of man are separate and unequal. Racial purity has been a core component of ideologies used to justify colonialism, slavery, eugenics, and various other forms of racial segregation and exclusion. The goal of maintaining racial purity is protecting whiteness. Racial purity as a tenet of white supremacy persists in contemporary racist ideology, including white nationalism. White nationalists have embraced the updated, geneticized version of white purity and polygenism. It's not surprising, then, that the term *racial purity* makes us flinch when used in polite company and mainstream discourse.

And yet belief in racial purity persists in the mainstream, as well. Practices used to sell DTC genetic ancestry testing dovetail with prevailing faith in genomic explanations for who and what we are and in the notion that our genomes encode our race(s) in discrete, quantifiable components. Practices used to categorize and market sperm deploy the terms *ethnicity* and *ancestry* as markers for race. The array of information blurs distinctions between the biographical, the genetic, and the socially constructed, so that every aspect of donor selection presents as genetic trait selection. Genetic ancestry testing and sperm bank companies characterize race as an elemental, insoluble component that can be measured, selected, and by implication, deselected. They sell racial purity.

The next section, "Distillation," defines racial purity and its role in the ideology of white supremacy. "The Emergence of Race" provides a selective history of explanations for race and racial purity, and their adaptations to the mid-twentieth century. "The Rise of Genetic Race" situates the production of genetic race alongside the formation of the biotechnology industry. "Racial Purity in the Market" examines the role of law and practices that AncestryDNA and California Cryobank use that instantiate genetic race, racial purity, and the new polygenism. The final section, "Genetic Identity," elaborates on commercial production of genetic identity in ways that draw from genetic race and its role in maintaining white supremacy, on the one hand, and from liberal discourses premised on the social construction of race, on the other. While these companies take no stance on racial politics, they sell concepts that serve no function outside of white supremacy.

DISTILLATION

Racial purity usually surfaces in literature about white supremacy or by white supremacists. When whiteness and purity are directly linked, we recognize racial purity as a racist idea. We know it when we see it presented that way. Yet racial purity is also embedded in everyday ways of thinking about identity. This chapter examines how racial purity functions in two markets premised on selling biological identity. This section sets the stage for that examination by pausing to consider the basic meanings instilled in racial purity.

Purity

Purity is a state of being untainted, uncontaminated, unmixed. Something pure is something elemental, consisting solely of one ingredient. We associate pure with true, clean, and natural. A pure heart. Pure motives. Pure can hone negatives, as well. Pure spite, pure greed, and pure hatred are concentrates, outside the range of governable emotions. In positive or negative form, that which is pure is unadulterated. An impurity is something that destroys the unadulterated state. Impurities found in water may ruin its quality. We often use synonyms for *not pure* or *impure* to cast aspersions. Things that are not pure are tainted, adulterated, or unnatural.

The simplicity of purity as a concept makes it useful as a vehicle for implied meaning, especially in value judgments. Moral belief systems, including the religious, use purity to confer certain actions or states of being with great virtue. Purity is the idealized state. Dictionary synonyms for *purity* include *chasteness, innocence*, and *immaculacy*.[2] You can imagine the antonyms.

Purity often conveys superiority relative to its opposite. Pure art and pure science hold themselves apart and above their commercial counterparts. Commercial art is art degraded by its use—to sell things. Commercial science is science driven by profit motive. Pure science is performed as knowledge seeking, which some regard as more morally worthy and less corrupted than commercial science.

Ironically, purity and its associated virtues have proved persuasive in commerce. Commercial advertising uses "purity" in taglines and name brands. Consider Ivory Soap, a name connoting whiteness for a Procter & Gamble product named in 1879. Within a few years, Ivory achieved fame and sales as the "safe, pure clean" body soap that floats. Its whiteness and buoyancy represent its lack of adulteration. Recent ads for Ivory Original Bar Soap include these highlights: "Free of dyes & heavy perfumes," "IT FLOATS," and "99.44% Pure."[3] Purity's appeal, in this context, is its association with nature. Ivory's message is that which is unadulterated is natural and superior to other soaps.

Racial Purity

The idea of racial purity starts with the assumption that racialized groups of people are distinct, determinable, and separable. It includes the claim that race in an unadulterated state can be attained. In addition, distilling race is not just possible but also meaningful. This, in turn, makes measuring or quantifying the content of one's race feasible, even necessary.

The concept of racial purity derives from the claim that race marks biological differences among human populations. Belief in a biological basis for racial difference persists despite the well-established fact that race is a social construct. Biological essentialism and the concept of racial purity sustain the persistence of belief in biological race. In other words, racial purity is a component part of belief in inherent racial difference. More specifically, racial purity is the idea that race can be distilled as an essential feature of a person or a population, and that each race can be distilled within a person or a population even after mixing has occurred.

The Purity of Whiteness

Racial purity is both core to the idea of biological race and foundational to claims of white superiority. The claim of white supremacy is possible only if the white race can be compared (favorably) with others and if race seems real. Mantling race in biology makes biological race appear to be both a neutral, proven claim and a natural feature of human life.

From its earliest days, biological race used phenotype to infer differences in physical, intellectual, and behavioral characteristics attributed to each racialized group. The methodologies used to produce evidence have changed over time. The ultimate goal—to justify racial white supremacist ideology—remains the same. For example, early constructions of race used physiognomy and ascribed character and intellectual profiles to explain the taxonomy of the five human types and the racial hierarchy.[4] In the early nineteenth century, the so-called science of race shifted to comparative anatomy, and to skull studies or phrenology in particular.[5] Natural history scholars and anatomists who studied phrenology explored the relationship between the shape and size of the human skull and behavior and intellectual capacity. The American race scientist Samuel George Morton, for example, used craniometry to produce evidence of inherent intellectual hierarchy among the races.[6] Phrenology is also notable because it relied heavily on measurement or craniometry. Craniometry expanded the use of quantification as a tool of establishing racial identity.[7]

Historically, the purity of whiteness mattered most. From its early days, white supremacy intertwined claims of inherent or natural racial hierarchy with strategies to protect the purity, and thus the supremacy, of whiteness. White supremacist ideology that valorized the purity of whiteness identified European forebears as the source of whiteness.[8] In short, within this ideology, 100 percent European ancestry makes one superior to those with lesser percentages.[9] This makes maintaining the purity of whiteness an explicit goal.[10]

Obvious and Nonobvious Racial Purity

Today, the association between white supremacist ideology and racial purity is both well understood and fraught, even—or perhaps especially—in globalized markets. In 2017 a company known for skin-care products, Nivea, launched a new ad that included the tagline "White Is Purity." The tagline appeared in a deodorant ad on Nivea's Middle East Facebook page. The ad prompted criticism of its racist messaging, while white supremacy organizations and individuals praised it. Mainstream media reported on the ad and the online discourse it prompted. Nivea pulled the ad two days later.[11] During that time, representative Facebook, Twitter, and other social media comments ranged from "We enthusiastically support this new direction your company is taking. I'm glad we can all agree that #WhiteIsPurity"[12] to "Not cool @NIVEAUSA @niveauk @NiveaAustralia . . . Not cool at all. #Racism is not a good marketing tactic."[13]

The ad's content and the public's response evidence the strength of the implied association between purity and whiteness, on the one hand, and white supremacist ideology, on the other. We are quick to recognize that association, even in a deodorant ad. While white supremacist organizations and those identifying as "alt-right" embraced the ad's "White Is Purity" line, the fact is that Nivea pulled the ad. Opposition to white purity messaging prevailed. And yet, in some contexts, we fail to recognize the use of racial purity or its white supremacist implications.

When "white" and "purity" are manifest, the association with white supremacy seems obvious. Without labels, the concept of racial purity is harder to detect. In fact, the concept of racial purity remains so deeply embedded in dominant culture and discourse that it implicitly shapes some liberal understandings of race, as well. People who describe themselves as one-half Black, one-quarter Asian, and one-quarter white may be using the categories to recount family history, pay tribute to their cultural affiliations, and celebrate their multiracial identity. And yet the quantification also echoes pernicious uses of racial purity. Dorothy Roberts observed, "we can only imagine someone to be a quarter European if we have a concept of someone who is 100 percent European."[14] Quantification recalls state laws that imposed racial classification based on the concept of blood quantum. Blood quantum rules have been used to classify people by race based on quantification of racial ancestry. More specifically for purposes of this discussion, states used blood quantum laws to determine whether the percentage of a person's nonwhiteness should affect their social and legal status, or their commercial value.[15]

Quantification suggests that race remains intact or insoluble even when mixed within a person or a group. Insolubility in this context does not deny that people from different racial groups may interact, form intimate relationships, and have children. Rather, insolubility conveys the belief that essential differences between races persist after individuals have overcome social barriers, as is evident in the concept of admixture discussed in the chapter by Carlos Andrés Barragán, Sivan Yair, and James Griesemer. Race mixing, abhorred by some and welcomed by others, is not inconsistent with belief in racial purity.

THE EMERGENCE OF RACE AND RACIAL PURITY

Race is not natural. Nor is it all that old. This section provides a brief account of race theories relevant to concepts that persist in twenty-first-century markets. Each version of race depends on racial purity. In the nineteenth century, two theories and assorted variations emerged. Monogenism, which asserts that all humans have a common ancestor, officially prevailed over its rival theory, polygenism.

Polygenism posited that the different racial populations emerged from distinct creation or evolution events. Although polygenism has become intellectually untenable, the idea of branched ancestral origins has persisted.

Race and Its Explanations

Explanations or theories for racial difference have changed over time. In general, theories that succeed in becoming influential use mantles of authority relevant to the era. The mantle, whether it be religion or science, validates the claim of racial difference as knowledge rather than mere belief. Yet politics have steered prevailing theory time after time. This discussion sketches how genetics emerged as the mantle of authority in theories of race and how law has implemented those theories and corollary concepts of racial purity.

The concept of race and its companion, racial difference, formed hand in hand with colonialism. Prior to colonialism, racialization did not occur.[16] As many historians have shown, empire was built on racial (and other forms) of subordination. These forces shaped colonial and early US law. Early colonial law in British North America defined racial categories and assigned racial identity. In the postcolonial United States, racialization continued and evolved. State and federal law incorporated and adapted colonial race laws.[17]

Colonial racial classification law protected the purity of whiteness. For example, a 1785 Virginia law defined as "mulatto" or mixed-race a person with at least one-quarter "Negro" blood. This law echoed a colonial-era ban on race mixing.[18] Later, more than a century after statehood, Virginia's racial classification law, like that of other states, set a more stringent standard. The law declared that "[e]veryone in whom there is ascertainable any Negro blood shall be deemed a colored person."[19] This version of state racial classification law came to be known as a "one-drop rule."[20] It zealously guarded white racial purity.

The legal definitions of *mulatto* and *colored person* relied on quantification and the claim that race is insoluble. You can mix Black and white, but the constituent parts remain intact. The use of fractions captures the insolubility of race. It also marks the limits of race: race mixing may upgrade Blackness in some contexts, but the person will remain less than white.

As the chapter by Meaghan O'Keefe will show, in early iterations of race, religion and science intertwined as a source of authority. According to the geneticist Joseph L. Graves Jr., "scientific ideology was not yet independent of Christian theology, and for this reason Western religion and science tended to be in general agreement concerning the significance and hierarchy of human races."[21] From the postcolonial era, science, religion, and combinations of both have persisted as mantles of authority.[22]

Monogenism and Polygenism

While theories of race proved adaptable over time, two macro theories or master narratives have competed for dominance. Monogenism asserts that there is one human species, originating from a common ancestral line. Proponents of polygenism believe that the human races have different origins and are therefore different species.[23] Each theory has its variations,[24] but for the purposes of this chapter, these versions suffice. In basic form, the two theories have served as templates or scaffolds for debates about the existence and salience of biologically based racial difference.

Both monogenism and polygenism have been used to assert that biology explains racial difference.[25] Early monogenists and polygenists set out "five separate human types: Caucasian, Ethiopian, Mongolian, American, and Malay."[26] Perhaps monogenists have had to work a little harder at it. For example, some monogenists claim that, while people of all races are of the same species, biological variation among racial populations is significant. They argue that long-term environmental pressures on populations located in different parts of the world produced those variations. Polygenism, on the other hand, aligns more easily with claims that racial difference and racial hierarchy are biologically inherent. Not coincidentally, polygenism ascended in the mid-nineteenth century, alongside defenses of slavery.[27] Some noted polygenists of the antebellum era made use of craniometry to link racial hierarchy, separate origins, and the immutability of race.[28] Each of those claims assumes that race can be distilled.

Officially, the debate among scientists and social scientists over the two theories lasted a relatively short time. As naturalists and biologists embraced Darwin's theory of evolution, monogenism prevailed over polygenism.[29] Yet, as discussed below, polygenist thinking persists in biomedicine. Belief in inherent racial difference, scaffolded by polygenist explanations, has also continued to shape racial discourse in society and law.[30]

Race and Nation

As noted, racialization arose hand in hand with colonization. Not surprisingly, then, theories of race extend beyond projects to classify and rank individuals and populations to defining national identity. Thus, the initial contest among European imperial powers over North America depended upon establishing the non-whiteness of Indigenous peoples. Once the fledgling US government formed, the relationship between nation and race became a continuing source of political tension. Countless examples illustrate this point, but consider, for now, the ideologies of eugenics and race suicide at the turn of the twentieth century.

In the late nineteenth century and the early twentieth century, many elite whites embraced the gene pool as a vehicle for social control by population control. Rationales for population control adapted select elements of the Mendelian genetic thesis.[31] Two overlapping ideologies proved appealing enough to effect legal change: eugenics and "race suicide."[32] Both prompted state legislatures and Congress to enact legislation aimed at controlling population growth vis-à-vis native-born whites.[33] Both also proved plastic enough to accommodate any number of groups targeted for social control.

Eugenicists ostensibly focused on the role of the so-called genetically fit and unfit. Eugenic goals included improving society's gene pool by encouraging pro-creation of the fit and preventing population increase of the unfit.[34] The most notorious eugenic strategy aimed directly at procreation.[35] States enacted laws that authorized involuntary sterilization of those deemed unfit. Statutory lists of those subject to forced sterilization varied widely.[36] Broad statutory interpretation prac-tices made it clear that poverty, breach of social norms (especially sexual mores), non-whiteness, and anything perceived as a disability could trigger the law.[37] US Supreme Court Justice Oliver Wendell Holmes Jr. validated Virginia's eugenic sterilization law:

> We have seen more than once that the public welfare may call upon the best citizens for their lives. It would be strange if it could not call upon those who already sap the strength of the State for these lesser sacrifices, often not felt to be such by those concerned, in order to prevent our being swamped with incompetence. It is better for all the world if, instead of waiting to execute degenerate offspring for crime or to let them starve for their imbecility, society can prevent those who are manifestly unfit from continuing their kind.[38]

As will be discussed at greater length in the chapter by Emily Klancher Merchant, eugenicists argued that improving the gene pool would benefit society and the nation, and that the resulting benefits justified the means.

"Race suicide" posited that the low birth rate among native-born whites rela-tive to non-whites and foreign-born whites would result in a society swamped by incompetence and moral decay.[39] Influential promoters of this thesis (including Theodore Roosevelt) called it the racial purity movement. They situated "race sui-cide" against a wave of immigration from China and southern and eastern Europe, and in the next few years, against the Great Migration. Calls for racial purity mea-sures ensued.[40] For some, the primary fear was the influx of Catholics, and concern they would outnumber Protestants. For many, the influx of groups deemed lower in status by ethnicity, race, and class made older measures like the Chinese Exclu-sion Act seem reasonable.[41]

Advocates of white superiority also touted scientific bases for eugenics.[42] Eugenicists and race suicide proponents supported population control laws such as marriage restrictions, including antimiscegenation laws, and immigration restrictions, as well as sterilization laws.[43] Both movements—eugenics and race suicide—deployed the so-called science of genetics and race to mobilize law and social policy against all but those deemed white, of northern and western Euro-pean descent, and fit. In short, the gene pool was used as a site to stake out a national identity based on race and class privilege.

The race suicide and eugenics movements were less coherent and less per-vasively accepted than this sketch suggests.[44] But the narratives that animated them reinforced the concept of biological race and the goal of racial purity. Both movements sought to engineer society through genetic control. More specifically, incorporation of genetic ideas strengthened the claim that race was a biological trait subject to measurement and quantification. Second, conclusions about fitness and unfitness—by disability, race, ethnicity, or other pseudotrait—conflated traits, social value, and moral capacity or lack thereof.

The Hardening of Heredity

Theories of race have shaped theories of heredity. The science historian Brad Hume has argued that nineteenth-century polygenists "hardened" heredity.[45] A "soft" theory of heredity posits that a combination of gene mixing and environmental influences produces a blend of acquired characteristics in a person. A "hard" theory of heredity sees heritable traits as fixed, resistant to environment, and persistent over time.[46] Within a hard theory of heredity, specific traits seen as characteristic of a race will remain intact, even if they skip a generation. Race-associated traits, then, act like some genetic disorders. This hard theory of heredity has itself remained intact in race theory.

THE RISE OF GENETIC RACE

State-sponsored eugenics lost ground in the 1930s and 1940s. Yet eugenic thinking and belief in biological race have persisted. As the science of genetics gained prominence, it became an influential platform for eugenic thinking and race theory. The most recent vehicle for biological race is genetic race. Genetic race has fueled the hardening of racialized heredity. Race theory, in other words, continues to adapt in the late twentieth and early twenty-first century.[47]

The Age of the Gene

A thumbnail sketch of genomics research often starts with the discovery in 1953 of the double helix structure of the DNA molecule, by Rosalind Franklin, Francis Crick, and James Watson.[48] This discovery enabled insight into what genes look like at the molecular level, how they replicate, and how they direct the chemical processes within cells. Within a short period of time, molecular biologists and other researchers generated new techniques, including the use of life's processes as lab tools, insights, and products. In the 1970s, this expanding body of work became the foundation of the biotechnology industry.

In the 1980s Congress jumped on the new genomics bandwagon. First came the Bayh-Dole Act of 1980.[49] Until this law became effective, patents on federally funded research remained under the government's control. The Bayh-Dole Act authorized academic, nonprofit, and small businesses to retain patent ownership and control of federally funded innovations. That act enabled institutions and researchers to commercialize their research, typically with industry partners. The Bayh-Dole Act effected significant change in biomedicine. It has spurred research institutions to use technology transfer to get biomedical inno-vations from bench to bedside and, thus, to produce revenue. In doing so, the act has also indirectly subsidized industry with the outcomes of federally funded research. Second, the Bayh-Dole Act made patents the coin of the biotechnology industry's rapidly expanding realm. Finally, the law effected a shift from biomed-ical research as a primarily public enterprise to a privatized one that positions patients as consumers. Genomics, a central activity of biotechnology, became a neoliberal enterprise.

In the mid-1980s conversations about a large-scale project to map the human genome began.[50] By 1988 Congress began increasing the federal budget for genome research.[51] The Human Genome Project (HGP) officially launched in 1990. Both the funding amounts and the descriptions cast the HGP as a big science project, akin to the race to the moon. President Bill Clinton and British prime minister Tony Blair announced completion of the draft map of the human genome in 2000. Both Clinton and Blair's speeches gave a hat tip to Watson and Crick (but not Franklin).[52] Both emphasized the enormous potential of the HGP to improve and save lives. Clinton, looking to the past, compared the HGP to Lewis and Clark's expedition. Looking to the future, he embraced privatization: "biotechnology companies are absolutely essential in this endeavor."[53]

Genetic Essentialism

The Human Genome Project produced two effects relevant to this analysis. First, it fostered genetic essentialism in research, medicine, and popular discourse. Genetic essentialism is the assumption that our genes provide the primary or exclusive explanation for health, illness, and even behavior.[54] A great deal of hype and hope accompanied the HGP. The metaphors used to describe the genome reflected the hype and hope. "Blueprint," "code," and "encyclopedia" of life spurred belief in the gene as the totalizing explanation for most aspects of human life.

Genetic essentialism valorizes the genome as the source code of why we are the way we are and who we are. It's a reductionist theory that in its simplest form posits that "there's a gene for that." Perhaps genetic essentialism's appeal is that it allows us to assume a one-to-one relationship between cause and effect, between gene and trait. Genetic essentialism focuses attention on molecular-level differ-ences within the body and then translates the hidden mechanisms into what we can see or think we can see. It takes the grade-school lessons we learned about Gregor Mendel's peas as the nearly exclusive way of thinking about who we are.

Genetic Race

Clinton's announcement in 2000 countered the idea that the genome codes for racial difference: "I believe one of the great truths to emerge from this triumphant expedition inside the human genome is that in genetic terms all human beings, regardless of race, are more than 99.9 percent the same."[55] Other official material

stated that humans are 99.9 percent genetically the same across racial populations.[56] Media coverage emphasized the finding.

And yet the Human Genome Project provided fodder for a resurgence of belief
in biological race.[57] It may be that biological race is so embedded in dominant culture that it filters and reconstructs what we hear. Perhaps the mention of race and
its colonial origins ("expedition") triggered that filter. Regardless, public and scientific discourses have either ignored the finding or mischaracterized the 0.1 percent
difference as racially significant. Since 2000 science and society have held the 0.1
percent accountable for genomic variations that justify claims of racial difference.

Genetic race is the updated version of biological race.[58] While genetic race deploys
new science,[59] it carries forward some of the old assumptions. Genetic race relies
on perceived associations between specific base pairs and their order and traits
associated with race. It carries phenotype inward, such that genes account not only
for phenotype but also for other racialized characteristics.

Genetic race has distorted research agendas, biotechnology markets,[60] health
policy, and health care. Biomedical research to determine the genetic bases for racial
differences in health, disability, and behavior gained credibility.[61] As seen in the
chapter by Tina Rulli and in the conclusion to this volume, medical providers have
felt justified in using racial profiling in delivery of health-care services.[62] Behavioral
genetics, which will be discussed at greater length in the chapter by Emily Klancher
Merchant, counts the founder of eugenics among its alumni and has proven ripe for
imputing racialization into its hypotheses, observations, and conclusions.[63]

The New Racial Purity

Mantling race in genomics may have strengthened the idea of racial purity. The logic
now goes something like this: As the building blocks of life, genes are the basic elements. The genes for race, then, are both elemental and insoluble. Gene sequences
for racialized characteristics are the distilled proof of race and racial difference.
This logic carries the thread of polygenism forward in time to the twenty-first century. As mentioned, polygenists forged a hard theory of heredity that constructed
traits as immutable and fixed. Contemporary use of "genetic ancestry," especially in
identity markets, describes ancestry in geographic terms. This practice echoes the
polygenist idea that differently racialized groups were geographically isolated and
must have evolved separately from each other. That hardened theory of race fits
within the dumbed-down geneticized version of biological race.

As Dorothy Roberts has shown, that logic leads to the conclusion that race-
associated diseases are genetic, deflecting attention from the role of structural racism.[64] Lower risk for breast cancer among Asians. Higher risk of diabetes in Latinx
people. High intelligence. Aggression. Placidity. It's all in their genes, insoluble,
unchangeable, and still bundled by race.

Twenty-first-century white supremacists have embraced genetic race and its
component parts, especially the purity of whiteness. News media and social media

have provided accounts of persons identifying as white nationalists using DTC genetic ancestry tests to prove their whiteness.[65] In part, white nationalists believe that maintaining white privilege and minimizing the presence and status of non-whites are core to what it means to be "American."[66] White nationalism promotes maintaining the purity of whiteness among white individuals and as a national identity. Using genetic ancestry test results to prove national belonging and ideological affiliation makes some sort of sense within that belief system.

Genetic race and racial purity are interlocking concepts, both contingent on genetic essentialism. Consider how this affects how we think about identity in the absence of white supremacy politics. In popular discourse, biocultural versions of race probably prevail.[67] That is, race in popular discourse mixes biological and cultural concepts. The biological concepts strongly shape how we talk and think about racial identity. For example, we assume that a person with an Asian forebear and a Black forebear is bound to receive a percentage of traits from each, respectively bundled as "Asian" and "Black." If each forebear is the person's parent, then the person is biracial, or half-Asian and half-Black. Half of her Black genes presumably remain intact as Black genes. The other half presumably remain intact as the genes for Asianness. While states no longer legislate "mulatto" classification or the one-drop rule, social norms still incorporate the practice of racial quantification that, in turn, animates racial purity.

RACIAL PURITY IN THE MARKET

The biotechnology industry and the Human Genome Project produced a swirl of research and discursive activity, often spurred by the hope and hype deployed to gain funding. As genetics emerged as the primary explanation for race and racial difference, genetic tools and use of human genes expanded. Genetic testing methods and uses have proliferated. Companies offer diagnostic testing, health risk assessment and prediction, and genetic ancestry description. The users and settings have also changed. Scientists use biotech tools in labs. Clinicians use them in medical settings. Other products are offered DTC as home-testing kits. In the meantime, collections of human cells, tissues, and DNA have become capital assets. Biobanks are curated for research, for therapy, and as collections of human data available not only for scientific discovery but also to commercial entities, consumers, and law enforcement. Well-known markets include human DNA biobanks, genetic testing, sperm banks, egg agencies, in vitro embryo banks for fertility purposes, and DTC genetic testing for medical and ancestry purposes.

Industries premised on DTC genetic ancestry testing and genetic selection are vehicles for social transmission of racial purity.[68] DTC genetic ancestry testing companies offer to provide genetic information to those who submit a sample of spit or other body materials containing DNA. Services include screening for genetic predisposition to everything from breast cancer to addiction to premature balding, carrier testing, paternity testing, noninvasive prenatal genetic testing, a child's potential for athletic or intellectual prowess, wellness information, and, of course, ancestry. The fertility industry not only uses genetic testing but also offers gametes and in vitro embryos for use with assisted insemination and in vitro fertilization. Both industries deploy practices that suggest and facilitate inference of connections between race, genes, and other traits. Both industries incorporate quantification methods that perpetuate the concept of racial purity.

The Law of Choice

The United States, relative to other countries, imposes little direct regulation of biotechnology markets. Generally, federal law provides a series of pathways to market, albeit with checkpoints. The Bayh-Dole Act, as discussed, promoted technology transfer and privatization of federally funded research work products. It expanded the role of patent and biobanking in biotechnology. Patent law standards and procedures, then, shape some aspects of the biotechnology markets. For the most part, patent law's stated purpose is to incentivize and reward innovation, without regard to necessity, efficacy, or social or ethical implications. Genetic ancestry testing methods and some other services they offer are, no doubt, patented. But patent law does not bar ethnicity estimates that instantiate genetic race and racial purity.

The US Food and Drug Administration (FDA) has authority to review and approve or disapprove for market a limited range of products. That authority includes human drugs and biological products and medical devices. Donated human semen is a biologic. The FDA does not, however, review ancestry tests.[69] When the FDA does review, it assesses clinical safety and efficacy of products. The agency can impose conditions on market distribution. As a result, sperm banks must register with the FDA. They also must obtain and review specified donor medical information and test for a specific set of communicable diseases.[70] But the FDA has imposed no conditions on how sperm banks curate and represent their product.

State law provides little to no direct restriction on sperm banks or genetic ancestry testing. Very generally, states tend to regulate assisted reproductive technology indirectly. State law consists largely of family law—to determine legal parentage when assisted insemination or in vitro fertilization have successful outcomes. State law regulation of ancestry testing is nonexistent or nearly so. In both sectors, the general laws of fraud, tort, or other consumer protection have the potential to redress some harms. But the companies carefully avoid offering facts or representations that are obviously actionable. Rather, their practices are crafted to invite conflation and interpretations structured by dominant discourse about race.

Privacy law in the United States is an ad hoc mix of federal and state law. Some state privacy laws address unauthorized disclosure of private information or failure to protect information by genetic ancestry testing companies or sperm banks.

But privacy does not really address the practices these companies use to produce race and reinforce racial purity.[71]

The US regulatory framework, such as it is, is notable for what it does not do. Other developed countries have regulatory approaches that screen products and new technologies to determine whether they should be developed or go to market. For example, comparative effectiveness research is used to compare harms, benefits, and costs of existing health interventions or products with new alternatives.[72] Arguably, comparative effectiveness assessment could be used on other technologies, as well. The United States, unlike Canada and much of Europe, rarely uses the precautionary principle, which aims to prevent or slow down new technologies that are potentially dangerous or have controversial social and ethical implications.

In contrast, the United States tends to allow evaluation of ethical, social, and even legal implications only when market distribution is inevitable or nearly so. Because those concerns have few, if any, legal handles, review of ethical and social implications is largely performative. In an industry founded hand in hand with neoliberalism, social norms impose limits based on consumer sensibilities. But in a society shaped by and inured to intense commercialization, concerns about commodification of human reproductive cells or racial identity have had only discrete force.[73] As a result, only minimal standards of good taste limit marketing messaging.

The absence of robust industry regulation and accountability places "personal responsibility," in neoliberal terms, on the consumer. The legal doctrine of informed consent serves this purpose beautifully. It presumes individual agency and validates placing the burden of protecting consumers on the consumer. The figure of the informed consumer, capable of determining exactly what she wants, backstops the lack of robust technology assessment.

In 1990 the California Supreme Court validated the assumption of agency in *Moore v. Regents of the University of California*.[74] John Moore sued his UCLA doctor, a researcher, the University of California, and its commercial partners. Moore had consented to a splenectomy two years before Congress enacted Bayh-Dole. Over the next seven years, he provided tissue samples for what his doctor said was necessary follow-up treatment for hairy cell leukemia. No one had mentioned using Moore's tissue and medical information for research and development of a cell line. When the case reached the California high court, his claims had been whittled down to two: breach of informed consent and conversion, a property-based tort. The court determined that John Moore had no property interest in his own cells and tissues, and therefore no claim for conversion. It did recognize a cause of action for breach of fiduciary duty or informed consent, but only against his doctor.

Moore v. Regents of the University of California serves as legal precedent only in California. But the case sets out the logic of acquisition that sperm banks and genetic ancestry testing companies use. Patients, sperm donors, and ancestry test users who submit spit samples effectively lose any property interest in their own cells and tissues once they leave the body. Sufficient disclosure confers protection against any other liability. Not surprisingly, *Moore* is the biotechnology industry's favorite case.[75] In *Moore*, informed consent documents with a sentence acknowledging the use of Moore's tissue for potential economic gain would have sufficed to protect the doctor. Upon disclosure, sperm banks and genetic ancestry testing companies can assert ownership of the cells and tissues. Sperm banks typically pay donors not because law requires purchase but to recruit inventory. The FDA requires medical screening, but otherwise sperm banks are free to market and sell to intended parents. Genetic ancestry testing companies have it better. They charge fees for providing genetic ancestry test reports to those who send spit samples, and if they disclosed other potential use and economic gain, they can also sell access to the information to third parties, subject to confidentiality protections.

Free market individualism reigns in the fertility and DTC genetic ancestry testing industries. Or rather, companies are free to market race, purity, and selection and to valorize individual choice. It's the vast unregulated spaces that law protects, rather than substantive regulation, that foster the production and purchase of genetic race.

Finding Ancestry, Making Race

The DTC genetic testing industry is global and growing. The North American market has the largest revenue share. Consumer use is expected to expand geometrically in the near future. In 2021 industry reports identified six or more segments in the DTC genetic testing market: "carrier testing, predictive testing, ancestry and relationship testing, nutrigenomics testing, skincare, and others."[76] Carrier testing and ancestry and relationship testing are the top two segments. 23andMe, Ancestry, and Color Health, Inc., consistently lead the industry.

AncestryDNA is the global leader in genetic relationship testing. "Know your world from the inside" appears at the top of the home page. Shortly below, the company website offers "your DNA story" based on DTC genetic testing.[77] The initial messages suggest that DNA contains everything you need to know about who and why you are. The claim that DNA provides a totalizing explanation taps directly into genetic essentialism.

The key to AncestryDNA's report is an "ethnicity estimate" that locates your genetic ancestors geographically.[78] Researchers challenge the methodology and content of the material that DTC ancestry testing offers.[79] This chapter focuses on specific aspects of the content. Social science definitions of ethnicity vary[80] but consistently use shared culture and identity as criteria. The use of geography depends on whether it informs shared group identity. In other words, ethnicity, like race, is socially constructed. In fact, the two are often conflated.[81] Ethnic identity arises from a sense of shared culture, heritage, sometimes language, and social experience.

In the United States, I have been assigned to and claim "Asian" as a racial category. Of course, others assign an identity to me that is a mix of ethnicity, race, and other social norms that have little to do with the ethnicities I claim. For example, like many others of Asian, Latinx, Middle Eastern, and North African descent, I am often cast as "foreign," "immigrant," and non-American. My assigned ethnicity also varies by time and place. When I first moved to Indiana in 1989, I was a presumptive Japanese foreigner. When media coverage of the 1992 civil unrest in Los Angeles hit the airwaves, I suddenly became a presumptive Korean. All of my grandparents immigrated from Japan, but I am not Japanese by ethnicity or nationality. Rather, depending on the context, my claimed ethnicity is Japanese American or Asian American, or sometimes Los Angeleno. Those socially constructed identities best fit my social experience within family, vis-à-vis dominant society and communities of color, including those I call my own. People whose grandparents immigrated from Japan to Cuba or France might have substantially different ethnicities. In other words, DNA cannot express ethnicity any better than it can express race.

Medical anthropologist Duana Fullwiley has told her personal experience of the social constructedness of race, in order to counter genetic race. "I am an African American," says Fullwiley, "but in parts of Africa, I am white." To do fieldwork as a medical anthropologist in Senegal, she says, "I take a plane to France, a seven- to eight-hour ride. My race changes as I cross the Atlantic. There, I say, '*Je suis noire*,' and they say, 'Oh, okay—*métisse*—you are mixed.' Then I fly another six to seven hours to Senegal, and I am white. In the space of a day, I can change from African American, to *métisse*, to *tubaab* [Wolof for "white/European"]."[82] AncestryDNA's "ethnicity estimate" is, at best, misnamed. Despite this, the website promises that as the company database grows, you will receive updates that correct the "ethnicity estimate."

The AncestryDNA website does not use the word *race*. It does link words such as *ethnicity, diversity*, and, of course, *ancestry*. Those words trigger consumer correlations between ethnicity and ancestry, on the one hand, and race, on the other.[83] In public discourse, race and ethnicity are often used in combination or interchangeably. Diversity and race are so often paired in public discourse that diversity must inevitably remind some viewers of race. As a result, geographic ancestry is conflated with race.[84] The website's images of people, family trees, and global maps also invite consumers to leap from ethnicity or ancestry to race. On AncestryDNA's website, many, if not most, of the photographic portraits are of people of color. The website's ethnicity lists include geographic regions like Oceania and the Balkans, countries like England and Norway, and names for racialized ethnic groups like Nilotic peoples and Maori. The elastic use of ethnicity provides space for interpolating race or simply conflating ethnicity with race. US consumers, embedded in culture and discourse that includes, for example, racial profiling of geographic regions, countries, and whole continents, readily interpret ethnicity estimates through the lens of race.

The website's message is that DNA, "cutting edge science," and "our science team" make all this possible.[85] The accompanying illustrations cluster photos of people of different phenotypes with labels for familial relationship, side by side with a representative ethnicity estimate that sums to 100 percent and a multicolored pie chart that presents the estimates in graphic form.[86] Thus, from genetic ancestry, race is readily distilled, quantified, and converted to separate colors, in the guise of science and technology-enabled precision. AncestryDNA's key product relies on the concept of racial purity.

While white nationalists have used genetic ancestry tests to prove the purity of their whiteness, others use the tests to affirm their multiracial identity. A study of 100,000 adults in the United States illustrates this point. Among other things, the study showed that people who identify as multiracial are more likely to have taken genetic ancestry tests.[87] It also concluded that those who take genetic ancestry tests "more frequently translate reported ancestral diversity into multi-racial self-identification."[88] AncestryDNA, in fact, promotes a geneticized version of diversity. The multicolored pie charts, world maps, and portraits suggest that racial diversity has been achieved—in biologized form.

What ancestry testing sells is a version of genetic race that has its roots in polygenism. The new polygenism does not insist that the races are different human species. But it assumes that race-specific genetic variation is significant enough to explain many differences among racial groups. This version of racial difference does not ostensibly premise white superiority. Racial purity, however, remains a core concept. This racially fractionalized version of identity also incorporates the hard theory of heredity. How else to explain the belief that racial identity is genetically represented in separable, insoluble percentages that sum to 100 percent?

Selecting Race, Making Descendants

Assisted reproductive technologies (ART) form the basis of a multibillion-dollar industry.[89] Core technologies include in vitro fertilization, assisted insemination, and egg freezing. People provide gametes—eggs and sperm—for others' use, in combination with assisted insemination, in vitro fertilization, and/or surrogacy. People who obtain others' gametes for their own use often do so through sperm banks and egg agencies. As discussed in the introduction to this volume, they are simultaneously acquiring a bundle of choices and a bundle of genes. Most consumers use ART to have a child with gametes from one or two intended parents, and thus to establish a genetic tie. Many intended parents use sperm and/or eggs that others provide, most often through sperm banks and egg agencies.

Industry analysts characterize the sperm bank industry by segments: semen analysis, storage, and donor. In the donor market, North America and Asia Pacific produced the largest revenues as of 2021.[90] The US market, in particular, has the highest revenue share. Of US-based sperm banks, California Cryobank is one of the largest in the domestic and global markets. While there are nonprofit sperm banks, most fertility businesses, like California Cryobank, are for-profit. Like its competitors, California Cryobank touts selectivity and sells gametic selection.

California Cryobank's website leads off its homepage with "Find Your Hidden Gem."[91] "Hidden Gems" is the name of "a carefully curated" portfolio of in-demand donors. The "Hidden Gems Gallery" contains donor numbers and photos suggesting why these donors are in demand. Most photos represent sports activities—soccer and basketball, for example. Others represent musical talent or professional achievement.

The website emphasizes the bank's selectivity in creating its catalogue of donors. Donors are described as "rare finds." The "Choosing Your Donor" page states: "California Cryobank's high standards and extensive screening process means our catalogue has nothing but the highest quality donors for you to choose from."[92] The Donor Recruitment page promises: "The majority of our sperm donors are recruited from world-class universities*, including UCLA, USC, Stanford University, Harvard University and MIT. Other donors are established professionals in various fields including business, medicine, law, and the entertainment industry."[93] And the "Donor Qualification" page opens with "Good Isn't Good Enough," followed by "[a]t California Cryobank our stringent donor qualification process allows less than 1% of all applicants to make it into our program."[94]

Messaging about selectivity and selection simultaneously anticipates consumer demand and shapes it. Basic qualification requirements for donors include a height minimum of five feet, nine inches, presumably because intended parents prefer tall donors.[95] In 2011 Cryos, one of the largest suppliers in the global sperm market, stopped accepting red-haired donors because it determined that its inventory was sufficient to meet limited demand.[96] Cryos officials explained that demand for ginger donors came only from Ireland.[97] Sperm banks also shape demand. California Cryobank, for example, provides a webpage and video under the heading "How To Find Your Perfect Sperm Donor." The information describes how to operate the digital catalogue. It also suggests selection criteria that align with the curated phenotype, medical history, and biographical profiles the company offers.[98]

Biographical and social achievement information about donors allows intended parents to find donors similar to an actual or imagined partner,[99] to satisfy hopes for a healthy or successful child, or to align with other values. For those using genetic selection to replace genetic descent, sperm selection offers a range of choices, packaged and priced for the discriminating consumer. The amount and detail of donor information that California Cryobank provides depends on the subscription level. California Cryobank offers three subscription levels, with the pitch that it's for your child. "Most likely, it's these little things that your child may find fascinating about your donor one day." The "little things" include whether donors described themselves as "artists, athletes, musicians, or scientists" and the childhood photos that enable "your son or daughter" to recognize "that button nose or big brown eyes as their own."[100] The pitch does not state that all donor characteristics are heritable, but intermixes those in which genetics play a role with those in which genetics do not.

On its "Donor Search" page, California Cryobank's website offers menus and access to donor profiles.[101] The mix of information presents intended parents a great deal of choice. Like the selectivity information, the selection information places donor information that may be genetic, and may even be heritable, alongside information that is biographical and not biological. A sample donor profile form intermingles phenotype descriptors, parental ancestry, high school and college GPAs, check boxes for mechanical skills and abilities, mathematical skills, sports played in high school or after, and language fluency. The last section of the form allows the donor to respond in their own words to queries such as hobbies and talents, how do you express your creativity, and what makes you laugh. Perhaps intended parents use the information to demedicalize a process that is an intimate one for people not using ART. Some intended parents construct a persona for the donor[102] in ways that reframe the act of shopping for gametes to something less commercial. Yet California Cryobank arrays that information in a format that suggests that donor selection is trait selection.[103]

The website does not include a menu labeled "race," although racialized choice is rampant in fertility markets.[104] Offers of racial selection use methods similar to AncestryDNA's ethnicity estimates. On California Cryobank's website, the Ethnic Origins and Ancestry lists are nonspecific and overlapping. Both contain racial categories and invite racialized readings of the information. The menu lists and donor profiles conflate race, country, and region. The Ethnic Origins list has seven items: American Indian or Alaska Native, Asian, Black or African American, Caucasian, East Indian, Hispanic or Latino, and Middle Eastern or Arabic. Most, if not all, of these items are constructed as racial and/or ethnic categories in the United States. The Self-Reported Ancestry list consists primarily of countries (the donor profile form prompts donors to identify countries in response to the Ancestry query). Notable exceptions include African American, American Indian, Caucasian, East Indian, Native American, and Native Canadian.[105] The interchangeable use of ethnicity, ancestry, region, and race simultaneously blurs the already fuzzy distinction between ancestry and race. Of the information deemed necessary to select a donor, "ethnicity" is third, along with medical history, height, GPA, and childhood photo.[106] The itemized list format for "ethnicity" reinforces assumptions that genetic race is both real and significant in donor selection. It also suggests that race remains discrete and fixed as components of the donor's body.

The company offers DNA Ancestry reports, along with the menu lists and donor profiles. The service offers intended parents the opportunity to "discover the biological ancestry for select donors."[107] Like AncestryDNA, DNA Ancestry provides estimates of "geographical ethnicity" in percentages that sum to 100 percent. The website claims the data is sufficient to provide "ancestry data for 26 unique geographic regions and ethnic groups," all color-coded.[108] Unlike Ancestry DNA, DNA Ancestry's use of ancestral origins is nearly exclusively (except Ashkenazi Jewish) a list of geographic origins, rather than a mix of geographic, racialized populations and ethnic-associated items. As discussed, while ancestral geographic origin may inform one's ethnicity, it's neither synonymous with nor determinative of ethnicity. The DNA Ancestry page explains why the company offers two types of ethnicity information. The text acknowledges that DNA Ancestry does not include "the donor's experiences and cultural identity," the type of information that social scientists consistently use to define ethnicity. The selling point is that "having both pieces of information can help create a more detailed picture of your donor to aid in donor selection."[109] In short, California Cryobank offers a carefully screened and curated set of choices, presented as traits and wrapped in color-coded percentages that sum to 100 percent.

Sperm banks like California Cryobank provide the opportunity to assemble racial identity, one composed of fractionalized components of race. Donors are the ancestors in the fertility industry. Of the many selections offered to consumers, race/ethnicity is prioritized. Other phenotyped features, biographical information, and medical screening data follow, as items bundled with "ancestry." Intended parents who choose the "selected donors" with DNA Ancestry reports double down on racial selection.

GENETIC IDENTITY MARKETS

Genetic ancestry test companies and sperm banks sell the opportunity to construct identity, attached to human tissue. Consumers of genetic ancestry test kits send spit samples and personal information. Companies like AncestryDNA then return a report, a bundle of information that consumers can use. Intended parents obtain reproductive material from sperm banks like California Cryobank after working their way through layers of choice, by which they gain access to a bundle of information about the donor. In both cases, the information, not the spit or semen, provides the means to construct identity based on twenty-first-century biological race.

In these markets, genetic race is a component part of the product. The new racial purity gives genetic race specificity. It makes fractionated identity, a thin representation of multicultural values, possible. It perpetuates the idea that race is insoluble and quantifiable. That old idea also helps sustain belief in polygenism. The new polygenism posits that genetic variations between races are significant and useful in research, health care, and kinship. The new polygenism incorporates monogenism by according less significance to the source of our species. In short, even if we can all trace our ultimate ancestors to one source, it's our racial ancestors that matter.

Both genetic ancestry testing and sperm bank companies offer services that increasingly tap into two technology sectors. During the past 30 years, makers of devices, tests, information banks, and an expanding range of products have made data about the self a technology sector and social phenomenon. Deborah Lupton calls this the "quantified self."[110] The quantified self, in Lupton's account, arises from self-tracking devices and the cultures formed around their use. Think Fitbit trackers or wearable sensors and "other computerised and automated ways of collecting personal information over a period of time."[111] DTC genetic testing stretches Lupton's technology boundaries, but seems apt in its use of quantified information presented with color-coded graphics that make the data digestible for nonexperts. It's the defining characteristics of racial purity—elemental, meaningful, and subject to precise measurement—that connect these identity markets. Like data produced by self-tracking devices, quantified race is shaping how we measure identity and imagine embodiment. It's not just race, but racial purity that sells.

Obviously consumers of genetic ancestry testing and sperm can accept or reject genetic race. They can use the bits and pieces that align with their preexisting sense of self. The bundles of information seem carefully assembled with enough space to permit individualized interpretation. At the same time, they direct use of genetic information in identity construction. White nationalists often interpret confounding results by deeming small fragments as insignificant. People who identify as multiracial are more likely to use genetic ancestry tests, and people who use genetic ancestry tests are more likely to identify as multiracial, despite the fact that the reports use "ancestry" and "ethnicity" and not "race." Some intended parents who are lesbians choose donors whose ethnicity and/or race differs from their self-identified race. Instead they prioritize the ability to use the same donor for future conceptions or to extend their already multiracial family identity.[112] In one case, family use of genetic ancestry tests revealed that decades earlier, a hospital had accidentally switched two babies. As a result, a person whose genetic family identified as white was raised in an Indigenous family and community, and the person with Indigenous ancestry was raised as white and with greater privilege. Both men reportedly faced uncomfortable, complicated questions about their identities. Both have recently stated that the test results do not change who they are, based on how they were raised, but they also feel a sense of loss.[113] Anecdotally, those statements are not singular. Others have also chosen their preexisting social and cultural identity over genetic identity.

These choices do not necessarily challenge the stability of genetic race. They may confirm that genetic race persists alongside the understanding that race is socially constructed. In the political flashpoint that race has become in the twenty-first century, the choice is between the two understandings of race. On the one hand, the Black Lives Matter movement has used the social construction of race to reveal how state law enforcement power masks violence against Black communities. The stark racial disparities in infection and mortality rates during the COVID-19 pandemic made undeniable the role of structural racism in health. Policy debates over use of race classifications in state law have prompted many states and the US Census to offer some flexibility in self-identification, including making limited versions of multiple race possible. And yet affirmative action opponents have produced state law that bans use of racial classification for education and employment purposes.

Genetic ancestry test companies and sperm banks are working the divide between the two theories of race. But, make of it what you might, what these companies sell maintains biological race, an updated version of polygenism—a theory inextricably grounded in defending slavery, and a new, perhaps hardened version of racial purity. At the same time, they foster—even celebrate—genetic multiracialism. The companies have no commitments to white supremacy. What they sell, however, has no neutral function. They are legacy concepts, adapted in twenty-first-century markets and hardened in twenty-first-century racial politics.

CONCLUSION: NEOLIBERAL IDENTITY

In a society where neoliberalism has prevailed, many aspects of our personal, even intimate, lives are governed through choice.[114] That is, our identities are partially formed in relation to commerce, through the exercise of free-market individualism. In identity markets based on genetic ancestry testing and sperm banking, companies offer genetic race and its components, racial purity and the new polygenism, in carefully curated, color-coded bundles. Free-market ideology says that consumers have freedom to use genetic race as they see fit. Yet market practices have preselected and refined the choices in ways that affirm the validity of genetic race and racial purity.

NOTES

1. For a rich discussion of genetic identity and race, see Keith Wailoo, Alondra Nelson, and Catherine Lee, eds., *Genetics and the Unsettled Past: The Collision of DNA, Race, and History* (New Brunswick, NJ: Rutgers University Press, 2012).

2. Merriam-Webster Dictionary, s.v. "purity," accessed July 1, 2023, https://www.merriamwebster.com/dictionary/purity.

3. "Ivory Original Bar Soap," Target, accessed November 12, 2023, https://www.target.com/p/ivory-original-bar-soap-10pk-3-17oz-each-it-floats/-/A-13951811.

4. Susan Branson, "Phrenology and the Science of Race in Antebellum America," *Early American Studies: An Interdisciplinary Journal* 15, no. 1 (2017): 164–93, 166.

5. Paul Wolff Mitchell, "The Fault in His Seeds: Lost Notes to the Case of Bias in Samuel George Morton's Cranial Race Science," *PLOS Biology* 16, no. 10 (October 4, 2018): e2007008, 2.

6. Ann Fabian, *Skull Collectors: Race, Science, and America's Unburied Dead* (Chicago: University of Chicago Press, 2010), 16; Mitchell, "Fault in His Seeds," 2, 9.

7. Branson, "Phrenology and the Science of Race," 167; Fabian, *Skull Collectors*, 177–78.

8. George Lipsitz, "The Possessive Investment in Whiteness: Racialized Social Democracy and the 'White' Problem in American Studies," *American Quarterly* 47, no. 3 (September 1995): 369–87, 371.

9. Aaron Panofsky and Joan Donovan, "Genetic Ancestry Testing among White Nationalists: From Identity Repair to Citizen Science," *Social Studies of Science* 49, no. 5 (July 2, 2019): 653–81, 656.

10. Robert Wald Sussman, *The Myth of Race: The Troubling Persistence of an Unscientific Idea* (Cambridge, MA: Harvard University Press, 2014), 38.

11. Amy B. Wang, "Nivea's 'White Is Purity' Ad Campaign Did Not Go Well," *Los Angeles Times*, April 5, 2017, https://latimes.com/business/la-fi-nivea-white-20170405-story.html.

12. Wang, "Nivea's 'White Is Purity' Ad Campaign."

13. "Nivea Removes 'White Is Purity' Deodorant Advert Branded 'Racist,'" BBC, April 4, 2017, https://bbc.com/news/world-europe-39489967.

14. Dorothy E. Roberts, *Fatal Invention: How Science, Politics, and Big Business Re-Create Race in the Twenty-First Century* (New York: The New Press, 2011), 228.

15. Judy Scales-Trent, "Racial Purity Laws in the United States and Nazi Germany: The Targeting Process," *Human Rights Quarterly* 259 (2001): 260–307, 282–84.

16. Joseph L. Graves Jr., *The Emperor's New Clothes: Biological Theories of Race at the Millennium* (New Brunswick, NJ: Rutgers University Press, 2001), 24–25.

17. A. Leon Higginbotham and Barbara A. Kopytoff, "Racial Purity and Interracial Sex in the Law of Colonial and Antebellum Virginia," *Georgetown Law Journal* 77 (1988): 1967–2029; Sharon M. Lee, "Racial Classifications in the U.S. Census: 1890–1990," *Ethnic and Racial Studies* 16, no. 1 (1993): 75–94, 77; Scales-Trent, "Racial Purity Laws," 282–84.

18. Winthrop D. Jordan, "Historical Origins of the One-Drop Racial Rule in the United States," *Journal of Critical Mixed Race Studies* 1, no. 1 (2014): 98–132, 112.

19. Scales-Trent, "Racial Purity Laws," 270.

20. Christine B. Hickman, "The Devil and the One Drop Rule: Racial Categories, African Americans, and the U.S. Census," *Michigan Law Review* 95, no. 5 (March 1997): 1161–265, 1187.

21. Graves, *Emperor's New Clothes*, 37.

22. Terence D Keel, "Religion, Polygenism and the Early Science of Human Origins," *History of the Human Sciences* 26, no. 2 (April 2013): 3–32, 5; Meaghan O'Keefe, this volume.

23. Joseph L. Graves Jr., "Favored Races in the Struggle for Life: Racism and the Speciation Concept," *Cold Spring Harbor Perspectives on Biology* 15, no. 8 (2023): 1–12, 3.

24. See O'Keefe, this volume.

25. Branson, "Phrenology and the Science of Race," 166.

26. Branson, "Phrenology and the Science of Race."

27. Graves, "Favored Races," 3.

28. Mitchell, "Fault in His Seeds," 3.

29. Graves, "Favored Races," 5–6.

30. Graves, *Emperor's New Clothes*, 45.

31. Brad D. Hume, "Quantifying Characters: Polygenist Anthropologists and the Hardening of Heredity," *Journal of the History of Biology* 41, no. 1 (2007): 119–58, 122; Alexandra Minna Stern, *Eugenic Nation: Faults and Frontiers of Better Breeding in Modern America* (Berkeley: University of California Press, 2005), 53.

32. Alexandra Minna Stern, "From 'Race Suicide' to 'White Extinction': White Nationalism, Nativism, and Eugenics over the Past Century," *Journal of American History* 109, no. 2 (2022): 348–61.

33. Miriam King and Steven Ruggles, "American Immigration, Fertility, and Race Suicide at the Turn of the Century," *Journal of Interdisciplinary History* 20, no. 3 (1990): 347–69, 348; Nancy Ordover, *American Eugenics: Race, Queer Anatomy, and the Science of Nationalism* (Minneapolis: University of Minnesota Press, 2003), 9–31; Warren S. Thompson, "Race Suicide in the United States," *Scientific Monthly* 5, no. 1 (1917): 22–35.

34. Daniel J. Kevles, *In the Name of Eugenics: Genetics and the Uses of Human Heredity* (Cambridge, MA: Harvard University Press, 1995).

35. Stern, "From 'Race Suicide' to 'White Extinction,'" 99–104.

36. Lisa C. Ikemoto, "Infertile by Force and Federal Complicity: The Story of *Relf v. Weinberger*," chap. 5 in *Women and the Law Stories*, ed. Elizabeth M. Schneider and Stephanie Wildman (New York: Foundation Press / Thomson Reuters, 2011).

37. See Paul Lombardo's *Three Generations, No Imbeciles: Eugenics, the Supreme Court and Buck v. Bell* (Baltimore: Johns Hopkins University Press, 2008), 58–63.

38. *Buck v. Bell*, 274 U.S. 200, 207 (1927).

39. Warren S. Thompson, "Race Suicide in the United States."

40. Thomas C. Leonard, "Eugenics and Economics in the Progressive Era," *Journal of Economic Perspectives* 19, no. 4 (2005): 207–24, 209–10.

41. Ordover, *American Eugenics*, 6.

42. Angela Saini, *Superior: The Return of Race Science* (Boston: Beacon Press, 2019), 82.

43. Kevles, *In the Name of Eugenics*, 96–112.

44. Kevles, *In the Name of Eugenics*, 129.

45. Hume, "Quantifying Characters."

46. Hume, "Quantifying Characters," 133.

47. Saini, *Superior*.

48. See Leslie A. Pray, "Discovery of DNA Structure and Function: Watson and Crick," *Scitable* 1, no. 1 (2008): 100, https://nature.com/scitable/topicpage/discovery-of-dna-structure-and-function -watson-397, for an account of the work that made the discovery possible.

49. Bayh-Dole Act. Pub. L. No. 96–517, § 6(a), 94 Stat. 3018.

50. Robert Mullan Cook-Deegan, "Origins of the Human Genome Project," *RISK: Health, Safety & Environment* 5, no. 2 (1994): 97–118, 102.

51. Cook-Deegan, "Origins of the Human Genome Project."

52. See "Text of the White House Statements on the Human Genome Project," *New York Times*, June 27, 2000, https://archive.nytimes.com/www.nytimes.com/library/national/science/062700sci -genome-text.html.

53. "Text of the White House Statements on the Human Genome Project."

54. Dorothy Nelkin and M. Susan Lindee, *The DNA Mystique: The Gene as a Cultural Icon* (Ann Arbor: University of Michigan Press, 1995).

55. "Text of the White House Statements on the Human Genome Project."

56. Catherine Bliss, *Race Decoded: The Genomic Fight for Social Justice* (Stanford, CA: Stanford University Press, 2012), 53.

57. Barbara Katz Rothman, *The Book of Life: A Personal and Ethical Guide to Race, Normality and the Human Gene Study* (Boston: Beacon Press, 2001), 107.

58. Roberts, *Fatal Invention*.

59. Catherine Bliss calls the use of genomics to sustain biological race "technologies of difference." Bliss, *Race Decoded*, 25.

60. Jonathan Kahn, *Race in a Bottle: The Story of BiDil and Racialized Medicine in a Post-Genomic Age* (New York: Columbia University Press, 2014).

61. Roberts, *Fatal Invention*; Bliss, *Race Decoded*; Kahn, *Race in a Bottle*; Theresa M. Duello, Shawna Rivedal, Colton Wickland, and Annika Weller, "Race and Genetics vs. 'Race' in Genetics: A Systematic Review of the Use of African Ancestry in Genetic Studies," *Evolution, Medicine, and Public Health* 9, no. 1 (2021): 232–45.

62. E.g., Sally Satel, "I Am a Racially Profiling Doctor," *New York Times*, May 5, 2002, https:// nytimes.com/2002/05/05/magazine/i-am-a-racially-profiling-doctor.html. For commentary, see Lisa C. Ikemoto, "Racial Disparities in Health Care and Cultural Competency," *Saint Louis University Law Journal* 48, no. 1 (2003): 92–93.

63. Ralph J. Greenspan, "The Origins of Behavioral Genetics," *Current Biology* 18, no. 5 (March 2008): 192–98.

64. Roberts, *Fatal Invention*.

65. Aaron Panofsky, Kushan Dasgupta, and Nicole Iturriaga, "How White Nationalists Mobilize Genetics: From Genetic Ancestry and Human Biodiversity to Counterscience and Metapolitics," *American Journal of Physical Anthropology* 175, no. 2 (2020): 387–98.

66. Leniece T. Davis, "Stranger in Mine Own House: Double-Consciousness and American Citizenship," in *Contemporary Patterns of Politics, Praxis, and Culture*, ed. Georgia A. Persons (New York: Routledge, 2005), 148–53; Abby L. Ferber, "The Construction of Race, Gender, and Class in White Supremacist Discourse," *Race, Gender and Class* 6, no. 3 (1999): 67–89, 81; Cheryl I. Harris, "Whiteness as Property," *Harvard Law Review* 106, no. 8 (1993): 1707–91, 1742–45.

67. Duncan Bell, *Dreamworlds of Race: Empire and the Utopian Destiny of Anglo-America* (Princeton, NJ: Princeton University Press, 2020), 34.

68. Laura Harrison, "The Woman or the Egg?: Race in Egg Donation and Surrogacy Databases," *Genders* 58 (Fall 2013): 24.

69. "What You Should Know—Reproductive Tissue Donation," US Food and Drug Administration, November 5, 2010, accessed February 8, 2024, https://www.fda.gov/vaccines-blood-biologics /safety-availability-biologics/what-you-should-know-reproductive-tissue-donation.

70. "What You Should Know—Reproductive Tissue Donation."

71. Many have heard of HIPAA (Health Insurance Portability and Accountability Act), the federal health information privacy law. HIPAA, however, has limited reach. It applies to "covered entities," which include health-care providers. But neither genetic ancestry testing companies nor sperm banks are "providers" under HIPAA. Most companies in these industries promise confidentiality. As a result, consumers harmed by breach of confidentiality may pursue a contract remedy.

72. John Donnelly, "Comparative Effectiveness Research," *Health Affairs Health Policy Brief*, October 5, 2010, https://www.healthaffairs.org/do/10.1377/hpb20101005.130478/full/healthpolicybrief _27-1555340068943.pdf.

73. Lisa C. Ikemoto, "Assisted Reproductive Technology Use among Neighbours: Commercialization Concerns in Canada and the United States, in the Global Context," in *Regulating Creation: The Law, Ethics, and Policy of Assisted Human Reproduction*, ed. Trudo Lemmens, Andrew Flavell Martin, Cheryl Milne, and Ian B. Lee (Toronto: Toronto University Press, 2017), 253–73.

74. *Moore v. Regents of Univ. of Cal.*, 51 Cal. 3d 120 (1990).

75. Kaushik Sunder Rajan, *Biocapital: The Constitution of Postgenomic Life* (Durham, NC: Duke University Press, 2006), 63–64.

76. See "Market Synopsis: Direct-to-Consumer Genetic Testing Market, by Test Type (Carrier Testing and Predictive Testing), by Technology (Targeted Analysis and Whole Genome Sequencing), and by Region Forecast to 2030," Emergen Research, September 2022, accessed February 8, 2024, https://emergenresearch.com/industry-report/direct-to-consumer-genetic-testing-market.

77. See homepage, AncestryDNA, accessed February 8, 2024, https://ancestry.com/dna.

78. "Your DNA Results, with More Detail than Ever," AncestryDNA, accessed February 8, 2024, https://ancestry.com/c/dna/ancestry-dna-ethnicity-estimate-update.

79. Deborah A. Bolnick, Duana Fullwiley, Troy Duster, Richard S. Cooper, Joan H. Fujimura, Jonathan Kahn, Jay S. Kaufman, Jonathan Marks, Ann Morning, Alondra Nelson, Pilar Ossorio, Jenny Reardon, Susan M. Reverby, and Kimberly TallBear, "The Science and Business of Genetic Ancestry Testing," *Science* 318, No. 5849 (2007): 399–400.

80. Chandra Ford and Nina T. Harawa, "A New Conceptualization of Ethnicity for Social Epidemiologic and Health Equity Research," *Social Science of Medicine* 71, No. 2 (2010): 251–58.

81. Ford and Harawa, "New Conceptualization of Ethnicity."

82. "Race in a Genetic World," *Harvard Magazine*, May–June 2008, https://harvardmagazine .com/2008/05/race-in-a-genetic-world-html.

83. Laura Mamo, *Queering Reproduction: Achieving Pregnancy in the Age of Technoscience* (Durham, NC: Duke University Press, 2007), 231.

84. Bolnick et al., "Science and Business of Genetic Ancestry Testing"; Sandra Soo-Jin Lee, Deborah A. Bolnick, Troy Duster, Pilar Ossorio, and Kimberly TallBear, "The Illusive Gold Standard in Genetic Ancestry Testing," *Science* 325, no. 5936 (2009): 38–39, 39.

85. Homepage, Ancestry, accessed February 8, 2024, https://ancestry.com.

86. DNA page, Ancestry, accessed February 8, 2024, https://ancestry.com/dna.

87. Sasha Johfre, Aliya Saperstein, and Jill A. Hollenbach, "Measuring Race and Ancestry in the Age of Genetic Testing," *Demography* 58, no. 3 (2021): 785–810, 793.

88. Johfre, Saperstein, and Hollenbach, "Measuring Race and Ancestry in the Age of Genetic Testing," 794.

89. Industry reports for 2022 assign a value for the US market of $8 billion. Five- to seven-year projections range from $16.8 billion to $46 billion, depending on the range of services included. See "US Fertility Clinics Market Report 2023," *Business Wire*, July 13, 2023, https://businesswire.com/news /home/20230713777238/en/US-Fertility-Clinics-Market-Report-2023-Sector-is-Expected-to-Reach -16.8-Billion-by-2028-at-a-CAGR-of-13.6; "Fertility Services Market Size Worth USD 46.06 Billion by 2030," *GlobeNewswire*, February 17, 2023, https://globenewswire.com/en/news-release/2023/02/17/2610675/0 /en/Fertility-Services-Market-Size-Worth-USD-46-06-Billion-by-2030-at-6-CAGR.

90. "Market Synopsis."

91. Homepage, California Cryobank, accessed February 8, 2024, https://cryobank.com.

92. "Choosing Your Donor," California Cryobank, accessed February 8, 2024, https://cryobank .com/how-it-works/choosing-your-donor.

93. "Donor Recruitment," California Cryobank, accessed February 8, 2024, https://cryobank.com /how-it-works/donor-recruitment.

94. "Donor Qualification," California Cryobank, accessed February 8, 2024, https://cryobank .com/how-it-works/donor-qualification.

95. "Donor Recruitment."

96. David W. Freeman, "Sperm Bank to Redheads: We Don't Want Your Semen," CBS News, September 10, 2011, https://cbsnews.com/news/sperm-bank-to-redheads-we-dont-want-your-semen.

97. Freeman, "Sperm Bank to Redheads."

98. "Choosing Your Donor."

99. Joanna Scheib, "The Psychology of Female Choice in the Context of Donor Insemination," in *Darwinian Feminism and Human Affairs* (New York: Springer Science & Business Media, 1997): 489–504, 497.

100. "Choosing Your Donor."

101. "Donor Search," California Cryobank, accessed February 8, 2024, https://cryobank.com /search.

102. Scheib, "Psychology of Female Choice," 499–500.

103. Harrison, "Woman or the Egg," 3.

104. Dov Fox, "Racial Classification in Assisted Reproduction," *Yale Law Journal* 118 (2009): 1844–98; Harrison, "Woman or the Egg"; Camisha A. Russell, *The Assisted Reproduction of Race* (Bloomington: Indiana University Press, 2018); Risa Cromer, *Conceiving Christian America: Embryo Adoption and Reproductive Politics* (New York: NYU Press, 2023), 144–53; Daisy Deomampo, "Racialized Commodities: Race and Value in Human Egg Donation," *Medical Anthropology* 38, no. 7 (February 7, 2019): 620–33; Jaya Keaney, "The Racializing Womb: Surrogacy and Epigenetic Kinship," *Science, Technology, and Human Values* 47, no. 6 (2019): 1157–79.

105. "Donor Search."

106. "Donor Information," California Cryobank, accessed February 8, 2024, https://cryobank .com/donor-search/donor-information/#Donor-Profiles.

107. "Donor Search."

108. "DNA Ancestry," California Cryobank, accessed February 8, 2024, https://cryobank.com /services/dna-ancestry.

109. "DNA Ancestry Page Sample Information," California Cryobank, accessed February 8, 2024, https://cryobank.com/_resources/pdf/sampleinformation/dnaancestrysample.pdf.

110. Deborah Lupton, *The Quantified Self: A Sociology of Self-Tracking* (Cambridge: Polity Press, 2016).

111. Lupton, *Quantified Self*, 2.

112. Alyssa M. Newman, "Mixing and Matching: Sperm Donor Selection for Interracial Lesbian Couples," *Medical Anthropology* 38, no. 8 (November 17, 2019): 710–24.

113. Norimitsu Onishi, "Switched at Birth, Two Canadians Discover Their Roots at 67," *New York Times*, August 2, 2023, https://www.nytimes.com/2023/08/02/world/canada/canada-men-switched-at-birth.html.

114. Jennifer M. Denbow, *Governed through Choice: Autonomy, Technology, and the Politics of Reproduction* (New York: NYU Press, 2015).

5

Reproducing Intelligence

Eugenics and Behavior Genetics Past and Present

Emily Klancher Merchant

In the early months of 2023, a thin, white, wealthy, bespectacled Pennsylvania couple began gracing the pages of newspapers and covers of magazines across the United States. Fearing that declining birth rates around the world would lead to what they termed "civilizational collapse," this couple—Malcolm and Simone Collins—had started the Pronatalist Foundation to encourage elite couples in wealthy countries to have more children.[1] Theirs is a high-tech pronatalism, advocating not just the use of assisted reproductive technologies but also polygenic embryo screening, a brand-new and yet unproven technique to identify the embryos in an in vitro batch with the lowest predicted risk of complex disease and the highest predicted capacity for mental health and educational success.[2] The term *pronatalism* refers to any effort to increase birth rates. The Collins' pronatalism, however, is more akin to positive eugenics—efforts to increase births only among a segment of the population considered superior—and in their case to choose superior embryos as well.[3]

While it is technically *possible* to assess the educational aptitude of an embryo, such screening is not commercially available, and scientists have argued that using this information to select an embryo for implantation would have little effect on the resulting child's actual educational attainment (compared to an embryo from the same biological parents selected at random).[4] Nonetheless, a 2023 survey found that nearly 40 percent of participants would strongly consider using predicted educational attainment as a basis on which to select their own embryos if such information were available at no cost.[5] Simone and Malcolm Collins used a DIY version of this screening for their third and fourth children.[6]

Writing about the Collinses in *Bloomberg*, Carey Goldberg says that "choosing your embryo based on its odds of earning a graduate degree is still a long way off from eugenics."[7] She is wrong. Eugenics is a scientific and political program first described in 1865 by the English polymath Francis Galton. He began with a policy proposal: that a range of social problems could be solved by breeding humans like livestock, selecting for socially desirable characteristics and against socially undesirable characteristics.[8] He then developed a scientific program that aimed to support selective breeding by demonstrating that mental and moral traits are primarily determined by biological material that is passed intact from generation to generation, what we now know as DNA.[9] In the pursuit of such evidence, Galton and his followers developed some of the fundamental tools of inferential statistics, tests for measuring intelligence, and methods for estimating the heritability of intelligence, or the proportion of variance in intelligence attributable to genetic variation.

Galton developed the concept of eugenics in England during a time when workers demanded the right to vote and when colonial subjects challenged imperial power in various parts of the world, most notably in the 1857 uprising against the British East India Company and the 1865 Morant Bay Rebellion in Jamaica. Galton claimed that the English class structure reflected variation in the biological inheritance of intelligence—those who had inherited more intelligence had higher positions in the social hierarchy—and that Britain ruled its empire because Europeans (and especially Anglo-Saxons) on average had more hereditary intelligence than did the non-white inhabitants of other continents.[10] His eugenic principles naturalized metropolitan socioeconomic inequality and imperial domination, and proposed a biological alternative to democratization and decolonization.

Although Galton's ideas did not get much traction initially, they began to catch on around the turn of the twentieth century. By the start of World War II, eugenics movements—also described in the chapters by Mark Fedyk, Lisa Ikemoto, and Meaghan O'Keefe—existed on every inhabited continent.[11] In the United States, eugenicists contended that Galton had shown the folly of the democratic project, disproving the claim that "all men are created equal."[12] Today, eugenics is often conflated with scientific racism. Scientific racists contended that members of different racially defined groups were not created equal. Eugenicists contended that even members of the same racially defined group were not created equal. Scientists established numerous eugenic organizations in the United States in the first decades of the twentieth century (many were established by the same people), conducting and promoting research on the inheritance of intelligence and other mental and moral qualities, and advocating for immigration restriction and involuntary sterilization.[13]

The word *eugenics* typically gets equated with policies regarding sterilization, immigration restriction, or genocide, but not with the scientific research that underpinned such policies. In the historical record, however, the two are impossible to separate. From Galton's day to the present, advocates of eugenic policies and programs have drawn on research into the measurement and inheritance of intelligence for support, and the scientists involved in that research have been among the most ardent proponents of eugenic policies and programs. They referred to their own science as eugenics and taught eugenics courses in universities.[14] Scientists' advocacy for eugenic policies might be understandable if the science of intelligence and its inheritance provided clear indications that differential intelligence is the primary driver of socioeconomic and racial inequality, and that differences in intelligence are primarily driven by genetic variation, but the science has always been inconclusive at best. Scientists today (in the 2020s) are only just beginning to figure out which genes might be involved in the development of human intelligence. Whether or how variations in those genes might produce different levels of intelligence from person to person (or group to group) remains unknown.

Eugenic policies and proposals, therefore, have always been *underdetermined* by the science. As this chapter will show, empirical evidence has never clearly supported scientists' claims, either that genetic variation is an important cause of social problems or that selective breeding could solve them. Instead, scientists' support for eugenic policies tends to shape the way they interpret and communicate their findings. In other words, the science—or at least its interpretation and communication—is often *overdetermined* by support for eugenic policies. Eugenic theory is a biological instantiation of racism and classism (the idea that socioeconomic and racial inequality inhere in the bodies of poor people and people of color rather than the structures of society) that long predates research into potential genetic causes of racial or socioeconomic differentials in intelligence. Such research, therefore, is subject to the influence of racism and classism at every stage of the process, from study design to communication of results. Advocacy for breeding programs is at the extreme end of eugenic policy proposals. Eugenic science has also underpinned advocacy against the redistribution of power and resources by suggesting that the existing order of things is natural and therefore changeable only through biological intervention or totalitarianism.[15] Science that claims to show a biological basis for existing racial and socioeconomic inequality therefore serves as a powerful antidemocratic force and deterrent to social change even in the absence of advocacy for selective breeding.

If selecting an embryo on the basis of its predicted educational potential doesn't *look* like eugenics to today's observers, that is because popular understandings of eugenics are overshadowed by the Holocaust. Discussions of eugenics frequently use the policies of the Third Reich as their benchmark, rather than the ideas of Francis Galton or the activities of the numerous eugenic organizations in the United States. As a result, they mistakenly reduce eugenics to genocide, race (pseudo)science, and state control over reproduction. But eugenics had a long and sordid history before and after the Holocaust, and it looked different from place to place. In the United States, it was remarkably flexible, adapting to shifting public opinion on racism, to developments in classical and molecular genetics, to the invention of assisted reproductive technologies, and to the rise of neoliberalism.

This chapter explores the long historical roots of recent research into the genomic correlates of education—the research that makes embryo selection possible. This research applies cutting-edge molecular methods to an older field of study, behavior genetics, whose history is intimately connected to that of eugenics. By tracing the institutional and intellectual relationship between behavior genetics and eugenics across the twentieth century and into the twenty-first, this chapter demonstrates that eugenics and behavior genetics pushed one another forward. Each advanced and responded to advances in the other, and made use of advances in assisted reproductive technology, even as many behavior geneticists began to distance themselves and their field from eugenics in the 1970s. The story focuses primarily on the United States, as behavior genetics inspired and received support from a version of eugenics that emerged in the United States in the 1930s and is intimately connected to the history of American race politics and the American civil rights movement.

Historians have identified the close relationship between eugenics and intelligence testing in the United States at the beginning of the twentieth century, demonstrating how eugenic principles shaped the development of intelligence testing and how the results of intelligence testing furthered eugenic projects.[16] The story typically ends, however, with the institutionalization of intelligence testing during and after World War I, and the use of wartime intelligence testing results to advocate for federal restrictions on immigration and the passage of state-level eugenic sterilization laws.[17]

This chapter continues the story, documenting how the eugenic aims of intelligence testers in the United States gave rise to the twin and adoption studies that transformed American eugenics and formed the core of behavior genetics until after the Human Genome Project. It also demonstrates that, as scientists developed more precise ways to measure the influence of DNA on intelligence and education—first through twin and adoption studies and more recently through genome-wide association studies—genetic influences have become less determinate and more elusive. Scientists still know very little about which genes may influence intelligence or education, and nothing at all about the biochemical mechanisms through which they may do so. Nonetheless, throughout this period, behavior geneticists have presented their research to the public *as if* it indicated a decisive role for genetics, and have advocated for policies premised on that overdrawn conclusion. The determinacy (and sometimes outright determinism) of scientists' public statements about the genetic causes of social outcomes is therefore at odds with the indeterminacy revealed by their own science, and this *indeterminate genetic determinism* has advanced a range of eugenic projects, from efforts to resegregate American public education in the 1960s to a sperm bank for Nobel Prize winners in the 1980s to polygenic embryo selection today.

INTELLIGENCE AND ITS HERITABILITY

Across the second half of the nineteenth century, Galton advocated for reproductive selection on a range of desirable characteristics. However, he often combined them into a conglomerate he termed "civic worth" and conflated with intelligence. Galton never developed an absolute metric for intelligence or civic worth; instead, he simply used socioeconomic status as a relative measure of it.[18] In fact, the first intelligence test was not developed for eugenic purposes. Created in 1905 by French psychologists Alfred Binet and Theodore Simon, the Binet-Simon test was designed to identify children who had fallen behind in school, so they could be given remedial education to help them catch up.[19] The test consisted of age-graded problem sets, designed so that approximately two-thirds to three-quarters of children of a particular age could solve the problems designated for that age.[20] Among other things, eight-year-olds were expected to be able to count down from twenty to zero, and nine-year-olds were expected to be able to name the months of the year in order.[21] The test measured things children were expected to have learned, not their innate capacity.

The meaning of the test changed, however, when it was imported to the United States by Henry Herbert Goddard, director of research at the Vineland Training School for Feeble-Minded Girls and Boys in New Jersey. Feeblemindedness was a central concept in American eugenics at the turn of the twentieth century. A catchall term describing those who deviated from the social norms of the day, it equated an unwillingness or inability to conform with substandard intelligence. Goddard presented the Binet-Simon test to his American colleagues as an objective tool to identify feebleminded individuals, not so they could receive remedial education, but so they could be prevented from spreading their feeblemindedness to future generations, either by institutionalization or by sterilization.[22]

Working closely with Charles Davenport, an American eugenicist who had collaborated with Galton in England, Goddard hired female fieldworkers to collect data on patterns of feeblemindedness in the families of Vineland children.[23] To manage these data, Davenport established the Eugenics Record Office (ERO) at Cold Spring Harbor, New York, in 1910 with a grant from the railroad heiress Mary Harriman. The ERO would eventually receive support from the Carnegie Institution for Science and the Rockefeller Foundation, two of the largest American philanthropies of the day.

By 1912 Goddard had collected enough data to publish a book titled *The Kallikak Family: A Study in the Heredity of Feeble-Mindedness*. The book told the story of Martin Kallikak, a pseudonym created from the Greek words *kallos* (beauty) and *kakos* (bad). Kallikak, Goddard claimed, was a Revolutionary War hero who had fathered two lines of descendants: one with his Quaker wife and the other with a "feebleminded" barmaid he had impregnated on his way home from the battlefield. According to Goddard, the descendants of Kallikak's wife were prosperous and intelligent, while the descendants of the barmaid were nearly all "feebleminded," with Kallikak's great-great-great-granddaughter ending up at Vineland and thereby coming to Goddard's attention.[24] The book became a national bestseller, popularizing eugenics for the first time in the United States.[25]

During World War I, Goddard teamed up with the Stanford University psychologist Lewis Terman to produce an intelligence test for US army recruits, evaluating over 1.7 million men before the armistice.[26] In the early years of the war, Terman revised the Binet-Simon test, renaming it the Stanford-Binet. Whereas the Binet-Simon, as used by Goddard, had primarily classified individuals as either feebleminded or normal, the Stanford-Binet drew on the concept of the intelligence quotient (IQ), introduced in 1912 by German psychologist William Stern, to produce a continuous measure of intelligence across the spectrum from low to high. Terman claimed that the test measured a person's innate capacity and therefore reflected their genetic value, or what Galton had termed "hereditary genius." Terman had explicitly eugenic aims for his test, predicting that it would "bring tens of thousands of these high-grade defectives under the surveillance and protection of society," which "will ultimately result in curtailing the reproduction of feeble-mindedness and in the elimination of an enormous amount of crime, pauperism, and industrial inefficiency."[27]

Results of army intelligence testing during World War I appeared to demonstrate a hereditary basis for the racial and socioeconomic inequality of the day. Following a pattern that could have been predicted by Galton, African Americans earned the lowest scores, followed by immigrants, with those from southern and eastern Europe earning lower scores than those from northern and western Europe. Native-born white men had the highest scores, but theirs were directly proportional to their socioeconomic status, with higher-class men receiving higher scores and lower-class men receiving lower scores.[28] Overall, more than half of American recruits had a mental age of fourteen or lower. Harry Laughlin, superintendent of the ERO, used these results to lobby for immigration restriction at the federal level and for eugenic sterilization laws at the state level.[29] Immigration restriction intensified in the mid-1920s, and 30 states adopted sterilization laws prior to World War II.[30] Over 33,000 Americans were sterilized under these laws between 1907 and 1939, with more sterilized after World War II.[31]

Just as Galton's eugenic theories had legitimated the restriction of democracy in Great Britain and the British Empire, Goddard, Terman, and other eugenic psychologists warned that most Americans did not have the innate intelligence required to participate in democratic self-government.[32] Intelligence tests had classified them as mental children, in need of superintendence by their supposedly natural superiors. Critics of these antidemocratic allegations, most prominently the journalist Walter Lippmann, countered that intelligence testing itself, not the low intelligence of the US population, posed the real threat to democracy.[33] Lippmann challenged Terman's key claims, first that a high IQ qualified one to lead society and second that IQ was inherited biologically.[34] Terman spent the rest of his life trying to prove the first point by following a cohort of high-IQ California children into adulthood.[35] These gifted girls and boys grew into amazingly accomplished women and men, though their success can't be attributed entirely to their IQ: Terman provided them with lifelong guidance, connections, and letters of recommendation.[36] Due to Terman's influence, a disproportionate number attended Stanford University.

Terman encouraged his students and other young educational psychologists to develop an answer to Lippmann's second critique by demonstrating that intelligence was inherited rather than acquired. This goal would prove elusive for Terman and continues to elude researchers today. Attempts to identify a genetic basis for intelligence built upon the modern evolutionary synthesis and a related statistical concept developed by the eugenic statistician Ronald A. Fisher, whom we met in the chapter by Mark Fedyk: the analysis of variance.[37] Theorizing that nature and nurture act independently to produce individual outcomes (which we now know is not true—nature and nurture are inextricably intertwined), Fisher contended that it was possible to measure the amount of variance in a trait in a sample that was caused by genetic (as opposed to environmental, or nongenetic) difference, a measure that, in the 1930s, came to be known as "heritability."[38] Heritability quickly became an important concept in animal husbandry, as it allowed breeders to estimate the effects of selective reproduction on future generations, under controlled environments. Eugenicists were interested in it for the same reason.

Animal researchers could estimate the heritability of given traits in given populations through breeding experiments, but educational psychologists could not. Instead, they adapted an analytic method developed by the animal geneticist Sewall Wright, known as path analysis. Path analysis allowed psychologists to decompose correlations between relatives in intelligence and other traits into genetic and environmental components by comparing sets of relatives with the same level of environmental similarity but different levels of genetic relatedness, such as adoptive parent-child pairs compared to biological parent-child pairs and monozygotic (identical) twin pairs compared to dizygotic (fraternal) twin pairs.[39] Terman edited the 1928 *Yearbook of the National Society for the Study of Education*, for which he solicited numerous path analytic studies of intelligence, hoping to establish, once and for all, that intelligence was inherited rather than acquired.[40] Yet these studies proved inconclusive. Each showed that intelligence was, in general, more tightly correlated among people who were more closely related, indicating some genetic influence. However, they did not definitively quantify the heritability of intelligence, and they indicated that nongenetic factors also play an important role in the development of intelligence. Terman nonetheless summarized these findings as evidence that a child's environment makes little difference to their intelligence. Regardless of environment, Terman concluded, "the feeble-minded remain feeble-minded, the dull remain dull, the average remain average, and the superior remain superior."[41] For Terman, these studies vindicated his assertion that intelligence tests provided an indication of innate genetic worth.

In the 1930s, however, psychologists would further challenge Terman's claim by demonstrating that IQ differences between Black and white Americans, and between US-born and non–US-born Americans, were driven largely by differences in home language and educational opportunities. In 1930 the Princeton University psychologist Carl Brigham, previously a strong proponent of northwest European superiority, admitted that his wartime findings on the genetic inferiority of southern and eastern European immigrants had been "without foundation." Further research had indicated that "comparative studies of various national and racial groups may not be made with existing tests," which penalized non-English speakers.[42] Beyond language, IQ tests relied on knowledge of and adherence to particular social norms. Terman had standardized the Stanford-Binet test on US-born white middle-class schoolchildren and adults in California, and many questions required cultural- and class-specific knowledge.[43] In 1935 two books by the psychologist Otto Klineberg attacked the contention that white Americans are innately more intelligent than Black Americans. Klineberg demonstrated that African Americans living in the North had higher IQ scores on average than white Americans living in the South, and that African Americans who moved from the South to the North showed greater gains in IQ with longer residence in the North.[44]

In the United States, intelligence testing and methods to estimate the heritability of intelligence were developed by adherents of eugenic ideology, who sought scientific evidence that intelligence was unequally distributed—both within and between groups defined by race and national origin—and that the distribution of intelligence was biologically determined. During the first decades of the twentieth century, when industrialization had produced immense socioeconomic inequality, intelligence testing and heritability studies generated apparent scientific evidence against social reform and in favor of selective reproduction and restrictions on democracy that facilitated selective reproduction. Although eugenics focused on biological explanations for *socioeconomic* inequality, it also undergirded a new scientific racism, one that looked to differences in average intelligence between groups as evidence of group-level superiority and inferiority.

Support for older forms of scientific racism began to wane at the end of the 1920s and the beginning of the 1930s, as scientists continually failed to find clear biological lines of demarcation between racially defined groups, and as race science became associated with the fascism emerging in Europe.[45] This turn away from scientific racism did not, however, signal the end of eugenics in the United States. In the 1930s, a new set of leaders at the American Eugenics Society (AES) rebranded eugenics. They developed a new eugenics program for the United States that was at least nominally free of the racism that was beginning to fall out of fashion and that minimized the state control over reproduction that was becoming a hallmark of European fascism.

REBRANDING EUGENICS

The AES was a relative latecomer to the eugenics scene in the United States, having been established only in 1926 by Charles Davenport, Harry Laughlin, and other eugenicists of their generation. It underwent a leadership transition in the 1930s. Older eugenicists, for whom eugenics had been inseparable from scientific racism, and who had focused their policy agenda on sterilization and immigration restriction, stepped down. Younger eugenicists stepped up, including Terman's former student and heritability researcher Barbara Burks. The most influential of these younger eugenicists was Frederick Henry Osborn, nephew of noted paleontologist and eugenicist Henry Fairfield Osborn, who had been a longtime president of the American Museum of Natural History and a founder of the AES.[46] Osborn, Burks, and their associates recognized that a eugenics program needed popular support in order to succeed in a democracy, and that popular support depended on scientific credibility.[47] They therefore created a new American eugenics program in the 1930s, one that reflected the current state of heritability research, jettisoned overt racism, and relied on market pressures rather than state power to influence birth rates.

The mission of the AES remained, as it had always been, "selecting the better and suppressing the poorer stocks."[48] Eugenicists of the older generation had understood race and national origin as indicators of supposed genetic quality. After all, the army intelligence tests had demonstrated that African Americans had lower intelligence scores than white Americans, and that foreign-born white men had lower intelligence scores than US-born white men. The younger eugenicists, however, argued that eugenic selection should be made on the basis of *individual* attributes rather than race or national origin. The attribute that was most salient to Osborn was a person's position in the socioeconomic hierarchy. He believed heritability studies provided good evidence that differences in intelligence between members of different socioeconomic strata were, at least to some extent, genetic in origin.[49]

Osborn did not, however, recommend that state or federal governments explicitly demand higher birth rates from higher-class couples or lower birth rates from lower-class couples. State control of reproduction was quickly becoming associated in the American popular imaginary with European fascism, and Osborn recognized that a successful eugenics program for the United States would need to be compatible with democracy. As noted in the chapter by Lisa Ikemoto, the Supreme Court had affirmed the constitutionality of eugenic sterilization in the 1927 opinion *Buck v. Bell*. Osborn, however, knew that the science of genetics was not yet developed enough to support a sterilization program that went beyond "carriers of severe defect."[50] The rest of the population would have to voluntarily have the number of children appropriate to their supposed level of genetic quality.

Osborn's proposed eugenics program therefore centered a set of social norms and financial incentives that would guide middle-class and wealthy couples to have more children and guide working-class and poor couples to have fewer. For wealthy couples, he expected tax breaks would encourage them to have more children. For middle-class couples, he recommended salaries proportional to family size and college scholarships for their children.[51] Osborn attributed large families among the poor to two things: ignorance of birth control and desperate conditions that undermined the initiative to use birth control. He therefore predicted that "better housing, and the improvement of economic conditions would bring a new sense of responsibility to the majority of these parents, and the extension of birth control knowledge, with new and cheaper methods of contraception, would then tend to reduce the proportion of very large families, and bring these groups below the replacement level," meaning fewer people in each successive generation.[52] Osborn did not expect that ameliorating the economic conditions of the poor would have any direct effect on improving society. Since he understood poverty to result from hereditarily low intelligence, he expected that real improvement would occur only through a reduction in family size among the poor, which would gradually take their genes out of circulation. He recognized that reducing the size of poor families without direct intervention would also necessitate the cultivation of new social norms, such as "a public opinion which will not tolerate families of more than one or two children among the socially inadequate, the dependent, the marginal economic."[53]

Although Osborn cited heritability research as evidence that socioeconomic status was a result of hereditary intelligence, heritability studies also demonstrated a role for the environment in the development of intelligence. They therefore generated popular support for efforts to improve the home and school environments of American children. To appease this sentiment, Osborn laid an environmentalist veneer on top of his hereditarian program. He emphasized that wealthier families provided better home environments for their children, agreeing with environmentalists "that the largest possible number of children should be brought up in the homes best fitted to develop their character and their intelligence, and the smallest possible proportion brought up by parents unable or unwilling to accept responsibility for such a home."[54] Osborn did not believe that these environments alone would increase intelligence, reduce poverty, or solve any other social problems, however. Rather, the environments proxied socioeconomic status, and therefore genetic quality, and Osborn believed that increasing the number of births to genetically superior parents and reducing the number of births to genetically inferior parents would increase intelligence in the aggregate and thereby ameliorate poverty.[55]

The 1930s therefore saw the emergence of a new brand of eugenics in the United States, one that is almost unrecognizable as eugenics if we take the policies of the Nazi government as our benchmark. Indeed, the proponents of this new American eugenics explicitly aimed to distinguish their program from the race-based and state-led policies that characterized German eugenics. The new leaders of the AES stopped talking about race, paid lip service to the role of home and school environments in the development of children's intelligence, and increasingly relied on individuals making market-based choices about the composition of their families. So what makes it eugenics? To begin with, its proponents called it eugenics, and called themselves eugenicists. They were the American *Eugenics* Society. More importantly, their program closely adhered to Galton's eugenic ideas and proposals, naturalizing socioeconomic inequality by presenting it as a result of genetic variation and proposing policies that would have enhanced the life chances of the middle class and the wealthy while diminishing those of the working class and the poor.

THE RISE OF BEHAVIOR GENETICS

Osborn believed that science would eventually prove the value of his proposals. Rather than waiting for science to catch up to eugenic theory, however, he helped it along by nurturing fledgling scientific subfields that he saw as potential allies for his eugenic project and whose practitioners needed support. In the 1930s, this was demography; in the 1950s, it was medical genetics and genetic counseling; and in the 1960s, it was behavior genetics, a subfield of psychology that aims to find genetic causes for human (and animal) behaviors and social outcomes.[56] Across the second half of the twentieth century, behavior genetics would lend valuable support to Osborn's eugenics program, generating apparent evidence that intelligence has a substantial genetic component, and that even the seemingly nongenetic influences on intelligence are themselves under genetic control. Behavior genetics also intersected with the backlash against the civil rights movement, opening a space for a new kind of scientific racism based in genetics.

By the beginning of the 1960s, Osborn had grown concerned that neither demographers nor geneticists were taking seriously the effects of changing birth rates on the intelligence of the American people.[57] He organized a series of conferences in Princeton, New Jersey, between 1964 and 1969 that aimed to put demographers and geneticists into conversation with one another. Over the years, the conferences drew in more and more psychologists working on the genetics of behavior, including Jerry Hirsch, Gardner Lindzey, John Loehlin, and Irving Gottesman.[58] These psychologists were the heirs to the research program on intelligence and its heritability that had been inaugurated by Lewis Terman and Barbara Burks in the 1920s. In 1970 they created the Behavior Genetics Association (BGA), with funding from the AES.[59] The two organizations remained close, connected by interlocking directorates.

Even before the BGA officially launched, however, the new field was thrown into controversy over the relationship between genetics, intelligence, and race. The prelude to the controversy was a 1967 publication in the *Proceedings of the National Academy of Sciences* by the UC Berkeley educational psychologist Arthur Jensen. Up to that point, psychologists had used a range of methods to estimate the heritability of intelligence.[60] There was no consensus about how the heritability of intelligence should be estimated, what the heritability of intelligence was, or what the heritability of intelligence *meant* beyond its technical definition.[61] Nobody could agree on what a high or low value of heritability was, or on what a high or low level of heritability indicated about the development of intelligence or its potential fixity or malleability. In the 1967 article, Jensen claimed to have answered these questions. He proposed a method that would become standard in the new field of behavior genetics for estimating the heritability of a trait in samples of monozygotic and dizygotic twins.[62] This method still produced a range of heritability estimates for intelligence, since heritability is a property of the sample in which it is measured, not a property of the trait itself. Jensen nonetheless announced that intelligence is 80 percent heritable, meaning that 80 percent of the variance in intelligence in a population is due to genetic variation.

Since heritability can range only from 0 to 1 (100 percent), a heritability of 80 percent, or 0.8, seems quite high. It is important to remember, however, what heritability means. It is an estimate of how much of the variance in a trait in a sample is due to genetic variance in the sample. It says nothing about how susceptible the trait is to change through environmental interventions. Jensen, however, claimed otherwise. He argued that a heritability of 0.8 meant that "if everyone inherited the same genotype for intelligence . . . but all non genetic environmental variance . . . remained as is, people would differ, on the average, by 8 IQ points." However, "if hereditary variance remained as is, but . . . *all* non genetic sources of individual differences were removed . . . , the average intellectual difference among people would be 16 IQ points."[63] Jensen therefore argued that the higher the heritability of a trait, the less it could be altered through environmental manipulation.

Jensen *must* have known that this interpretation was simply untrue, as a 1958 study in rats had clearly demonstrated that genotype and environment are not independent of one another: the amount of difference genes make depends on the environment, and the amount of difference the environment makes depends on genes.[64] There is therefore no way to say how much variance there would be under a fixed environment, or how much variance there would be under a fixed genotype, without specific information about the environment or the genotype. In other words, the numbers Jensen provided for these hypothetical scenarios were pure speculation. He nonetheless announced that "these results decidedly contradict the popular notion that the environment is of predominant importance as a cause of individual differences in measured intelligence in our present society."[65] Other scholars in the emergent field of behavior genetics would have known that Jensen's conclusions were unwarranted. Publishing in *PNAS*, however, allowed Jensen to get away with these misleading claims. As a high-profile general science journal, its audience likely would not have known enough about the genetics of behavior to do anything other than take Jensen at his word.

Jensen's claims about the biological fixity of intelligence served a larger political purpose that became clear in 1969, when the controversy began in earnest. In an article published in the *Harvard Educational Review* (another nonspecialist journal), Jensen presented the high heritability of intelligence as evidence that programs like Head Start would never close the IQ gap between Black and white students in the United States because the gap was rooted in genetic difference.[66] Jensen called for the resegregation of American education, and for a eugenics program that would reduce the childbearing of all individuals with low IQs, which would have disproportionately targeted African Americans, given racial bias in IQ testing.

The Nobel Prize–winning physicist William Shockley had been using his scientific celebrity status to advance similar claims for a few years by that point, and Jensen's article seemed to add the scientific authority that Shockley lacked because he didn't have a background in genetics.[67] The two men had met during Jensen's sabbatical at the Center for Advanced Study in the Behavioral Sciences at Stanford University in 1966–67, and both received support from the openly racist Pioneer Fund, whose explicit goal was to reinstate educational segregation in the United States after the Supreme Court's 1954 decision in *Brown v. Board of Education*.[68]

Geneticists in the 1960s knew that Jensen's and Shockley's claims for a genetic basis to average IQ differences between Black and white Americans had no foundation in heritability studies or any other scientific evidence.[69] Heritability estimates refer only to the proportion of variance *within* a sample that is due to genetic variation; they can say nothing about the cause of differences *between* samples. As the population geneticist Richard Lewontin explained, "the fundamental error of Jensen's argument is to confuse heritability of a character within a population with heritability of the difference between two populations." This was a problem because, according to Lewontin, "between two populations, the concept of heritability of their difference is meaningless."[70] At the end of the 1960s, the heritability of intelligence had been estimated only in white Americans and Europeans. Such estimates provided no evidence regarding the source of average IQ differences between Black and white Americans or any relative genetic superiority or inferiority for either group vis-à-vis the other. Indeed, there was—and still is—no scientific method to assess the role of genetics in producing group-level differences in IQ or any other trait. Given the structural racism that has always plagued the United States, it is just as plausible that African Americans have the superior genetics, but that these are overwhelmed by an environment of severe oppression.[71]

In support of his racist claims, Jensen merely pointed to his 0.8 heritability estimate, arguing that it showed environment to play little role at all in the development of intelligence; he claimed that average differences between racially defined groups therefore *must* have at least some genetic component. Lewontin pointed out in numerous scientific and public forums that Jensen was simply wrong: even if the heritability of intelligence among white Americans was 1, or 100 percent (essentially meaning that the environment made no contribution to differences in intelligence between white Americans), this would still say nothing about the causes of average differences in intelligence between Black and white Americans.[72]

Other scientists argued that Jensen had overestimated the heritability of intelligence. This overestimate had occurred in three ways. First, the data Jensen had drawn from studies of identical twins reared apart were simply bogus. In some studies, the data appear to have been fabricated.[73] In all of the others, the phrase "reared apart" was interpreted so loosely as to be nearly meaningless.[74] Second, the method of estimating heritability by comparing samples of monozygotic twins to samples of dizygotic twins, which Jensen had presented as a new gold standard, was known at the time to overestimate heritability, both because monozygotic twins tend to grow up in more similar environments than dizygotic twins, and because monozygotic twins share *all* of their DNA—including interaction effects between genes (epistasis)—so they are actually more than twice as similar genetically as fraternal twins. For these reasons, the animal geneticist Douglas Falconer had explained in 1960 that a comparison between monozygotic and dizygotic twins can produce only an "upper limit" to estimates of heritability[75]—that is, an overestimate.

The third way in which Jensen overestimated heritability was that his method attributed to genetics "any variance attributable to the interaction of genotype and environment,"[76] including genes that had no direct bearing on intelligence but that shaped a person's social world in ways that might influence their intelligence. Education scholar Christopher "Sandy" Jencks explained what this meant in colloquial terms in 1972:

> If, for example, a nation refuses to send children with red hair to school, the genes that cause red hair can be said to lower reading scores. This does not tell us that children with red hair cannot learn to read. Attributing redheads' illiteracy to their genes would probably strike most readers as absurd under these circumstances. Yet that is precisely what traditional methods of estimating heritability do. If an individual's genotype affects his environment, for whatever rational or irrational reason, and if this in turn affects his cognitive development, conventional methods of estimating heritability automatically attribute the entire effect to genes and none to environment.[77]

While Jensen and other behavior geneticists were (and still are) happy to include this type of "genetic cause" in their heritability estimates (because it makes intelligence seem more "genetic"), it does not represent what most people think of when they imagine potential genetic effects on intelligence or education.[78] Behavior genetics thus engages in a type of reasoning that is directly opposed to feminist theory, critical race theory, and disability studies, each of which separates social and somatic causes of inequality. Each of these liberatory approaches attributes inequality to discrimination, not to the bodies of the people being discriminated against. Behavior genetics does the opposite, presenting the effects of discrimination as originating in an individual's DNA. While feminist, antiracist, and disability scholars work toward dismantling discrimination by denaturalizing inequality, behavior genetics promotes discrimination by naturalizing inequality.

Many nonscientists reacted with outrage to Jensen's racism. Protesters disrupted his lectures and threatened physical harm. The tires on his car were slashed, and police had to open his mail. Jensen received bomb threats at his office, and his family had to seek protection.[79] This response allowed Jensen to portray himself as a victim, even as he advocated genocide against African Americans according to the UN definition of the term, which includes restricting births among a racially or ethnically defined group.[80] The public focus on Jensen's racism centered race differences in IQ in the popular debate, leaving unquestioned whether IQ had any practical significance. Galton and Terman had proposed that intelligence directly determined a person's socioeconomic status and value to society, but sociologists in the 1960s had found that *educational attainment* was the key to socioeconomic success in the United States, and that intelligence was not the sole determinant of educational attainment; a child's parents' socioeconomic status mattered at least as much.[81]

Jensen's supporters compared him and other behavior geneticists advancing racist claims to Galileo, a truth teller being persecuted by irrational zealots. The BGA, and the field of behavior genetics in general, rallied around him. As behavior geneticists defended Jensen, they became hyperfocused on estimating the heritability of mental traits and behaviors using methods similar to the one Jensen had described in 1967.[82] These studies suggested that *all* traits and behaviors are heritable, though heritability estimates varied wildly between samples for the same trait.[83] They also appeared to show that social institutions—such as families, schools, and religion—played only a trivial role in individual outcomes.[84] Echoing Frederick Osborn, behavior geneticists claimed that a child's home environment was genetically determined, influenced by the genes of both parents and children. Even the amount of television a child watched, it seemed, was heritable.[85] In the epistemological space of behavior genetics, heritability created a kind of hall of mirrors from which there was no escape. Genes seemingly accounted for *all* social outcomes, though the methods that appeared to demonstrate this supposed fact provided no information about how any actual genetic variants might influence any of them.

Behavior geneticists reiterated Jensen's misleading statements about the meaning of heritability estimates and defended his "intellectual freedom" to make scientifically unwarranted claims about the relationship between race and intelligence.[86]

To these white and mostly male scientists, protecting Jensen's freedom to speculate idly about the innate inferiority of an oppressed segment of society was more important than protecting his targets from the consequences of such speculation. An attempt by the wider genetics community—the Genetics Society of America (GSA)—to make a clear statement to the American public that "there is no convincing evidence of genetic difference in intelligence between races" failed because several GSA members insisted that it would be equally true to say that "there is no convincing evidence that there are *not* genetic differences in intelligence between races."[87] Ultimately, the GSA took a nonposition on the issue, stating that "in our view, there is no convincing evidence as to whether there is or is not an appreciable genetic difference in intelligence between races."[88]

As behavior geneticists doubled down on their support for Jensen, the gulf between behavior genetics and other social sciences widened.[89] Researchers outside of behavior genetics put little stock in heritability studies, so behavior geneticists developed their own publishing ecosystem to bring their work into print. They published in eugenics journals, many of which were in the process of taking the word *eugenics* out of their titles (such as *Annals of Eugenics*, which became *Annals of Human Genetics* in 1954; *Eugenics Quarterly*, which became *Social Biology* in 1968 and is now *Biodemography and Social Biology*; and *Eugenics Review*, which became *Biosocial Science* in 1969). They also published in new journals specific to behavior genetics (such as *Behavior Genetics, Twin Research, Intelligence*, and *Personality and Individual Differences*). There was even a set of journals (such as *Mankind Quarterly; Journal of Social, Political, and Economic Studies*; and *Population and Environment*) for research that was too racist to appear in the other journals.[90]

Those who did this racist research received generous support from the Pioneer Fund. When Richard Herrnstein and Charles Murray published *The Bell Curve* in 1994, they disproportionately cited scholars who had received support from the Pioneer Fund and whose work was published in *Mankind Quarterly*. Herrnstein and Murray's argument differed little from the one advanced by Jensen and Shockley in the 1960s. Publishing 25 years later, however, they could make the disingenuous and obviously untrue claim that the civil rights movement had equalized opportunities between Black and white Americans, so any remaining disparities in IQ or socioeconomic status "must" be genetic in origin.[91] In response to widespread criticism of the book, 52 behavior geneticists, many of them Pioneer Fund grantees, published an open letter in the *Wall Street Journal* in Herrnstein and Murray's defense. Titled "Mainstream Science on Intelligence," the letter portrayed the book as having been based in solid scientific research.[92] The claims it made were considered "mainstream" only among behavior geneticists, but the letter's publication in the *Wall Street Journal* elevated those claims to the status of established fact among the American public. Similar ideas were also aired in other popular press outlets, such as *Science News*, which in 2022 apologized for its earlier support for eugenics and scientific racism.[93]

Around the same time, behavior genetics authorized a bizarre eugenic venture. In 1980 the Repository for Germinal Choice opened just outside San Diego. One of the country's first sperm banks, it offered the gametes of Nobel Prize–winning (male) scientists to high-IQ women, who could presumably use them to have smarter children than they would be able to conceive with their male partners.[94] Few Nobel Prize winners ever donated their sperm—William Shockley was the only one who publicly admitted to having done so—and the repository eventually cast a wider net, trawling the halls of university math and science departments and targeting self-made businessmen.[95] Though it went out of business just before the turn of the millennium, the repository created a new consumer-focused model of sperm donation that has only gained in popularity since then, as described in the introduction to this volume and the chapter by Lisa Ikemoto.

The repository's legitimacy depended on the indeterminate genetic determinism that formed the heart of both Osborn's eugenics program and the field of behavior genetics. Men who donated sperm to the repository did not undergo any kind of genetic testing. Since behavior genetics had demonstrated the heritability of intelligence, the Nobel Prize itself served as a genetic marker. As sperm banking grew in popularity, choosing a donor at least partly on the basis of his test scores or educational attainment became the norm, demonstrating general public acceptance of eugenic principles grounded in the indeterminate determinism of behavior genetics.[96]

During the last few decades of the twentieth century, the meaning of eugenics shifted yet again. Jensen, Shockley, and the Pioneer Fund used the word *eugenics* to describe their explicitly racist breeding proposals. A new organization, the American Eugenics Party, sprang up in the mid-1960s, vocally equating eugenics with racism.[97] It seemed that Osborn and the AES had lost the 30-year battle to divorce eugenics from racism in the popular imaginary. In 1972 the organization changed its name to the Society for the Study of Social Biology.[98] Its program remained the same, but its leaders, now primarily drawn from the new field of behavior genetics, wanted to distance the organization from the word *eugenics*, which was no longer separable from racism. Ironically, the behavior geneticists associated with the erstwhile AES were among the less racist members of their field.

As the leaders of the organization embraced the new name, they also projected it backward in time, reinterpreting the previous 30 years of the organization's history. In this revisionist version, eugenics had never changed; the organization had simply stopped doing eugenics around the time of World War II. The 1990s saw an outpouring of histories of eugenics, covering most parts of the world. The majority of this scholarship ended before 1945, producing the popular impression that eugenics had ended then as well.[99] Osborn's eugenics was no longer eugenics; it was now simply behavior genetics, medical genetics, genetic counseling, and fertility medicine. This rewriting allowed behavior geneticists to disavow and forget the eugenic origins of their field, even as some continued to hail Francis Galton as its founder.[100] It also reduced eugenics to racism, genocide, and state control over reproduction, making it impossible to recognize or critique such eugenic initiatives as the Repository for Germinal Choice and the polygenic screening of embryos for educational potential because they aren't explicitly racist and they operate on the private market rather than through the state.

GOING MOLECULAR

By the turn of the twenty-first century, behavior genetics had demonstrated that all human outcomes are heritable but had produced no information about which genes might contribute to which outcomes or how they might do so. Some behavior geneticists continued to point to heritability estimates as evidence that average IQ differences between racially defined groups were genetic in origin, while others maintained that heritability demonstrated no such thing. The field had exhausted the limits of the twin method popularized by Jensen in 1967. The indeterminate determinism of behavior genetics underpinned sweeping claims: that the existing social order was rooted in genetic difference and therefore natural, just, and immutable; that most findings in sociology and economics were wrong because they didn't take genetics into account; and that racial inequality was a product of genetic difference rather than discrimination. At the beginning of the twenty-first century, behavior genetics went molecular.[101]

After the completion of the Human Genome Project, it began to seem possible that behavior geneticists might finally overcome their field's indeterminacy by locating the actual genes that contribute to intelligence and socioeconomic status. Other social scientists also became interested in genetics at this point. Sociologists and epidemiologists were excited to identify the genes that predispose people to complex diseases in order to better tease out the social causes.[102] Some sociologists were also curious about the genetics of behavior.[103] In the quantitative social sciences, outcomes are always underdetermined, meaning that, no matter how many variables a model includes, it will never be able to account for all or even most of the variance in the outcome. Sociologists suspected that genes might explain why people in the same social circumstances often respond in different ways.[104]

Behavior geneticists and their new partners initially looked for correlations between specific traits and genes with known biochemical effects. Within a decade, however, it became clear that this candidate-gene approach wasn't working. Researchers attained few positive results, and even fewer of these replicated. The most well-known is probably the so-called "warrior gene," a variant of the MAOA gene that was found to predispose men to aggressive behavior. When this result failed replication, behavior geneticists hypothesized that perhaps it caused aggression only in people who had been abused as children.[105] Further research, however, showed that children who were abused were more likely to grow into aggressive adults regardless of which variant of MAOA they possess.[106] Nonetheless, Genex Diagnostics still sells an over-the-counter test for the "warrior gene."

In 2012 a group of genetically oriented social scientists announced that "most reported genetic associations with general intelligence are probably false positives."[107] This finding didn't shake behavior geneticists' faith that intelligence was driven largely by DNA, but it did encourage them to adopt a new paradigm. In keeping with the modern evolutionary synthesis, behavior geneticists had long worked on the assumption that intelligence and socioeconomic status were polygenic—that is, influenced by multiple genes. This assumption didn't change, but after the failure of candidate-gene studies, behavior geneticists decided that, instead of looking for a small number of genes with large effects, they should look for a large number of genes with tiny effects.[108] They termed this idea the "fourth law of behavior genetics."[109]

Following the lead of medical and psychiatric genetics, behavior geneticists and their new collaborators in economics and sociology turned to genome-wide association studies. Known familiarly as GWAS, these hypothesis-free studies simultaneously but independently test millions of loci (single-nucleotide polymorphisms, or SNPs) across the genome for correlations with the outcome in question. Since they seek minuscule effects, they require enormous samples. The Social Science Genetic Association Consortium (SSGAC) was born in 2012 from the need for these huge samples. As a consortium, it can meta-analyze cohorts across a variety of studies to get the statistical power necessary to identify tiny genetic effects. But it was difficult to do a GWAS on intelligence, as most genetic studies hadn't tested participants' IQ, and those that had done so had used a variety of different metrics. Nearly all of the available data sources, however, had collected information about participants' educational attainment, which became the SSGAC's primary outcome of interest. Over the past 10 years, the vast majority of research in molecular behavior genetics has focused on educational attainment.

The SSGAC published its first GWAS of educational attainment in 2013.[110] Although the study would prove highly consequential, its findings were not particularly impressive. It identified three SNPs with statistically significant correlations to educational attainment, each of which was associated with about a month of additional schooling. When summed into a polygenic score—which molecular behavior geneticists describe as an index of a person's genomic propensity for a particular outcome (in this case, educational attainment)—DNA appeared to account for only about 2 percent of the variance in educational attainment, leaving 98 percent unexplained by genetics. Because the study used cutting-edge molecular methods, and because it was published in *Science*, arguably the highest-profile outlet for scientific research, it generated a new respectability for behavior genetics, even though the findings were meager and even though the idea that educational attainment has a genetic basis sounds preposterous to most people.

The popular press reported on the study with an appropriate level of skepticism. *Futurity* stated that "genes have small effect on length of education."[111] The *Chronicle of Higher Education* announced that "there is no gene for finishing college."[112] Even the *Wall Street Journal* cautioned readers that there probably isn't a "gene for" height or intelligence.[113] Those closer to the study, however, read its results differently. The SSGAC's leadership believed that a GWAS run on a larger sample could produce a polygenic score that accounted for *more* than 2 percent of the variance in educational attainment. They were right. The SSGAC published two more studies of educational attainment, in 2016 and 2018, the latter using a discovery sample of 1.1 million people and generating a polygenic score that accounted for approximately 12 percent of the variance in educational attainment.[114] Behavior geneticists and their new colleagues responded to the 2016 and 2018 studies with breathless enthusiasm, publishing books for popular audiences that touted GWAS and the polygenic scores they generated as a validation of the genetic determinism represented in twin and adoption studies.[115] A fourth GWAS came out in 2022.[116] With a sample size of 3.3 million, it managed to raise the proportion of variance accounted for up to 16 percent, as shown in figure 5.1. At the same time, however, the study showed that the majority of this effect was predictive but not causal. At most, it appears that only about 5 percent of the variance in educational attainment can be attributed to the *causal* effects of DNA.[117] This is a far cry from the 40 percent heritability estimated for educational attainment from twin studies.[118] Rather than suggesting that twin studies may have overestimated heritability, however, behavior geneticists argued that they simply needed different methods to find the genes responsible for the "missing heritability."[119]

In addition to being small, molecular research shows that the effects of DNA are largely drowned out by those of childhood socioeconomic status. In a study of older white Americans, individuals with the *highest* polygenic scores for educational attainment but whose fathers were in the bottom quartile of the income distribution were less likely to have graduated from high school and college than were individuals with the *lowest* polygenic scores but whose fathers were in the top quartile of the income distribution.[120] Similarly, white kids with low polygenic scores for educational attainment are more likely to complete advanced math classes in high school if they attend wealthy schools than if they attend poor ones.[121]

A serious problem with molecular behavior genetics is that it includes only white people.[122] This is true of most GWAS, as discussed in the chapter by Tina Rulli, not just GWAS for social or behavioral outcomes. These studies typically use supposedly "ancestrally homogenous" samples to avoid spurious associations, and they typically define "genetic ancestry" in continental terms.[123] This practice conflates genetic difference (which varies continuously across space) with US race categories (which identify people categorically according to the migration history of their ancestors), furthering the popular but incorrect belief that race categories represent genetic difference. It also produces faulty results. Researchers have

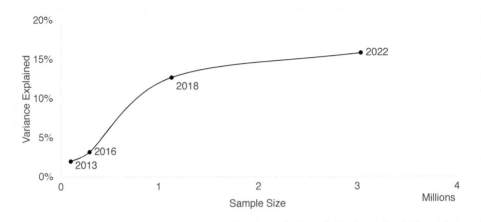

FIGURE 5.1. Genome-wide association studies (GWAS) of educational attainment. The *x* axis shows the size of the discovery sample; the *y* axis shows the proportion of variance accounted for by the resulting polygenic score. As the size of discovery samples increased, so, too, did the variance accounted for by the resulting polygenic scores, but further increases in the size of the discovery sample will likely have diminishing returns. Image created by the author.

found that the racial exclusivity of medical GWAS threatens to exacerbate health disparities,[124] and the same would undoubtedly be true if the GWAS for educational attainment were used for educational or policy purposes. Molecular behavior geneticists have largely brushed this problem aside, claiming that GWAS will become more representative any day now.[125] While it is true that initiatives like the Human Pangenome Reference and the National Institutes of Health's "All of Us" project are increasing the diversity of genome databases, much work still needs to be done to overcome the technical challenges to performing GWAS on genetically diverse samples. Until then, research in molecular behavior genetics will be limited primarily to white people, and research has demonstrated that polygenic scores are more predictive for some white people than for others.[126]

Molecular behavior geneticists and their colleagues are well aware of these limitations and have published at length about them in venues frequented by specialists. But they present a very different image in venues intended for popular audiences. To be sure, most molecular behavior geneticists do not write for popular audiences. Those who do, however, routinely oversell the role of genetics in producing social outcomes and exaggerate how much we know about the role genetics plays in producing social outcomes. In public-facing publications, scientists misrepresent the findings of behavior genetics research—including their own research—to claim that genomic variation makes a decisive contribution to differences in intelligence, educational attainment, and socioeconomic status.[127] At times, popular descriptions of polygenic scores for educational attainment and other socioeconomic outcomes equate them with "genes for" the outcomes they predict, and at other times as measures of the outcomes themselves.[128] Their authors describe polygenic scores as valuable tools for social scientific research, personalized educational interventions, and public policy.

Molecular behavior geneticists who write for popular audiences represent a tiny fraction of their field but serve as ambassadors to the general public, not just in the United States but also worldwide. As such, they foster the widespread acceptance of deterministic ideas about the effects of DNA on behavior, even when they themselves disclaim genetic determinism.[129] Many present themselves as political progressives. Nonetheless, their research has been used for a range of reactionary, eugenic, and racist purposes. The behavior geneticist Robert Plomin has argued that the polygenic score for educational attainment should be used to allocate educational opportunities and occupational placements, describing it as a test of intelligence and aptitude that people can neither cheat on nor study for.[130] In 2015 the sociologist Dalton Conley—a coauthor of the 2013 GWAS of educational attainment—published a popular online article describing how the polygenic score for educational attainment could be used for embryo selection.[131] Although his vision was decidedly dystopian, it represented polygenic embryo selection as *effective* and may therefore have inspired readers like Simone and Malcolm Collins.

Polygenic scores are more determinate than the heritability estimates produced by twin and adoption studies in the sense that they provide individual predictions of the outcomes for which they are constructed, though the SSGAC has warned against using polygenic scores for educational attainment in this way. Polygenic scores are, however, still indeterminate in the sense that they provide no information about which variants *contribute* to the outcome in question (as opposed to simply predicting it) or how they do so. Variants contributing to educational attainment *might* make people more intelligent, but they might just make them taller or more attractive, such that other people respond to them in ways that encourage them to go farther in school. Overall, however, they simply contribute to the indeterminate genetic determinism of modern eugenics, producing more concrete evidence that genes matter in some way without producing any information about how.

Molecular methods, therefore, have considerably boosted the authority of behavior genetics without advancing scientific knowledge about how DNA might contribute to either intelligence or education. Scientists now have a sense of which genomic variants correlate with educational attainment in white people with supposedly European genetic ancestry, but they have also recognized that correlation is not the same as causation, and they are still no closer to identifying biochemical mechanisms that might link DNA to education or any social outcome. Nonetheless, behavior geneticists have widely publicized GWAS and the polygenic scores they produce as validation of the eugenic idea that intelligence and socioeconomic status have a genetic basis. Such hype inspired a New Jersey start-up, Genomic Prediction, to make it possible for IVF patients to screen embryos for "intellectual disability," the company's disingenuous gloss for low predicted educational attainment, though this service was quietly discontinued at the end of 2020 due to bad press.

CONCLUSION: EUGENICS TODAY

When prospective parents like Simone and Malcolm Collins select embryos on the basis of their predicted educational attainment, they may not be aware that they are participating in eugenics. They are certainly not engaging in genocide, and the government is not selecting their embryos for them. The racism of the endeavor is hidden from view—companies that sell polygenic embryo screening do not advertise the fact that the science behind their product was carried out on white people and that polygenic scores are far more predictive for white people than for people of color.[132] Such parents are also likely unaware that the science behind the polygenic score for educational attainment is indeterminate at best. Since 2016 behavior geneticists have presented GWAS and polygenic scores to the public as if they demonstrated a decisive role for genetics in educational attainment, playing down the fact that polygenic scores explain very little of the variance in educational attainment and the fact that any biochemical mechanisms that *might* connect DNA to educational attainment remain completely unknown. Since behavior geneticists have obscured their field's long roots in eugenics, today's prospective parents are likely unaware that the GWAS for educational attainment is simply the most technologically advanced approach in a eugenic research project that originated with Galton's desire to breed humans like livestock. This research agenda has produced no information about which genes might contribute to the development of human intelligence or how they might do so, but has produced widespread acceptance of the idea that intelligence is largely under genetic control, that white people have more of the "genes for" intelligence than people of color, that the existing socioeconomic hierarchy is natural, and that social interventions can do little to change it.[133]

The real problem with eugenics is not that the Collinses and their followers will actually be able to breed smarter children. As noted above, scientists have found that embryos selected on the basis of their polygenic score for educational attainment would be unlikely to attain much more education than a randomly selected embryo from the same biological parents.[134] Rather, it is that attributing socioeconomic inequality to genetic diversity is simply the wrong diagnosis, one that ignores a century of scientific, historical, and genetic research. As such, it can only point to ineffective or at best inefficient solutions that are more likely to perpetuate inequality than to overcome it. Eugenics doesn't "work" by breeding better people; it works by convincing us that socioeconomic and racial inequalities are underpinned by biological variation, and that some people are therefore more deserving—of education, wealth, power, rights, and even life—than others.

It works by absolving governments, social institutions, and individuals from the responsibility of improving the world we all share.

The most chilling consequence of the SSGAC's research agenda probably could have been foreseen in advance. Just as Arthur Jensen, William Shockley, Richard Herrnstein, and Charles Murray called on heritability studies to advance the racist claims that African Americans have a lower genetic endowment of intelligence than white Americans, today's race scientists have pointed to the results of educational GWAS to make the same racist claims.[135] Although GWAS of educational attainment have been done only on white people, and although molecular behavior geneticists have warned against drawing any kind of racial comparisons on their basis, white nationalists have pointed to their results to make unsubstantiated assertions that African Americans have fewer of the intelligence- and education-producing variants than white Americans.[136] The results have been nothing short of devastating. In 2022 a white supremacist cited the SSGAC's third GWAS of educational attainment in a racist diatribe he posted shortly before perpetrating a mass shooting at a grocery store in an African American neighborhood in Buffalo, New York.[137] While the SSGAC is certainly not responsible for this heinous act of violence, it underscores how easy it is to unwittingly promote racism, inequality, and even genocide when we do not understand the history of eugenics and thereby fail to recognize the eugenic projects in which we may be participating.

NOTES

1. Simone Collins and Malcolm Collins, "Why the World Needs More Big Families like Ours amid the Population Crisis," *New York Post*, January 28, 2023, https://nypost.com/2023/01/28/the-world -needs-more-big-families-like-ours-for-humans-to-survive.

2. Carey Goldberg, "The Pandora's Box of Embryo Testing Is Officially Open," *Bloomberg*, May 26, 2022, https://www.bloomberg.com/news/features/2022-05-26/dna-testing-for-embryos-promises-to -predict-genetic-diseases.

3. Arwa Mahdawi, "'Hipster Eugenics': Why Is the Media Cosying Up to People Who Want to Build a Super Race?," *Guardian*, April 21, 2023, https://theguardian.com/lifeandstyle/2023/apr/20 /pro-natalism-babies-global-population-genetics.

4. Genomic Prediction currently scores IVF embryos for lifetime risk of a variety of complex diseases. When the company first opened its doors in 2019, the information it provided to prospective parents included an indication of whether an embryo was predicted to result in a child with an "intellectual disability" (code for low predicted educational attainment), but this service was quietly discontinued at the end of 2020. Currently, a prospective parent who uses Genomic Prediction's services can download their raw data and upload it to one of a number of other websites that will evaluate it for educational potential. For critiques of the efficacy of embryo selection on this basis, see Patrick Turley, Michelle N. Meyer, Nancy Wang, David Cesarini, Evelynn Hammonds, Alicia R. Martin, Benjamin M. Neale, Heidi L. Rehm, Louise Wilkins-Haug, Daniel J. Benjamin, Steven Hyman, David Laibson, and Peter M. Visscher, "Problems with Using Polygenic Scores to Select Embryos," *New England Journal of Medicine* 385 (2021): 78–86.

5. Michelle N. Meyer, Tammy Tan, Daniel J. Benjamin, David Laibson, and Patrick Turley, "Public Views on Polygenic Screening of Embryos," *Science* 379, no. 6632 (February 9, 2023): 541–43.

6. Jenny Kleeman, "America's Premier Pronatalists on Having 'Tons of Kids' to Save the World: 'There Are Going to Be Countries of Old People Starving to Death,'" *Guardian*, May 25, 2024, https://www.theguardian.com/lifeandstyle/article/2024/may/25/american-pronatalists-malcolm-and-simone-collins.

7. Goldberg, "Pandora's Box of Embryo Testing."

8. Francis Galton, "Hereditary Talent and Character," *Macmillan's Magazine* 12, nos. 68 and 70 (June and August 1865): 157–66 and 318–27.

9. Ruth Schwartz Cowan, "Nature and Nurture: The Interplay of Biology and Politics in the Work of Francis Galton," *Studies in History of Biology* 1 (1977): 133–208.

10. Francis Galton, *Hereditary Genius* (London: Macmillan, 1869); Francis Galton, *Essays in Eugenics* (London: Eugenics Education Society, 1909).

11. Alison Bashford and Philippa Levine, eds., *The Oxford Handbook of the History of Eugenics* (New York: Oxford University Press, 2010).

12. See, for example, Madison Grant, *The Passing of the Great Race, or the Racial Basis of European History* (New York: Charles Scribner's Sons, 1916).

13. Alexandra Minna Stern, *Eugenic Nation: The Faults and Frontiers of Better Breeding in America* (Berkeley: University of California Press, 2005).

14. Paul Popenoe and Roswell Hill Johnson, *Applied Eugenics* (New York: Macmillan, 1918).

15. See, for example, Toby Young, "The Fall of the Meritocracy," *Quadrant Online*, September 7, 2015, https://quadrant.org.au/magazine/2015/09/fall-meritocracy.

16. Stern, *Eugenic Nation*.

17. John S. Carson, *The Measure of Merit: Talents, Intelligence, and Inequality in the French and American Republics, 1750–1940* (Princeton, NJ: Princeton University Press, 2007); Nancy Ordover, *American Eugenics: Race, Queer Anatomy, and the Science of Nationalism* (Minneapolis: University of Minnesota Press, 2003).

18. Galton, *Essays in Eugenics*.

19. Carson, *Measure of Merit*.

20. Lewis Madison Terman, *The Measurement of Intelligence* (Boston: Houghton Mifflin, 1916), 37.

21. Terman, *Measurement of Intelligence*, 38.

22. Leila Zenderland, *Measuring Minds: Henry Herbert Goddard and the Origins of American Intelligence Testing* (New York: Cambridge University Press, 1998).

23. Correspondence with Henry Herbert Goddard, box 41, Charles B. Davenport Papers, American Philosophical Society, Philadelphia.

24. Henry Herbert Goddard, *The Kallikak Family: A Study in the Heredity of Feeble-Mindedness* (New York: Macmillan, 1912).

25. J. David Smith and Michael L. Wehmeyer, "Who Was Deborah Kallikak?," *Intellectual and Developmental Disabilities* 50, no. 2 (2012): 169–78; Stephen J. Gould, *The Mismeasure of Man* (New York: W. W. Norton, 1981).

26. John S. Carson, "Army Alpha, Army Brass, and the Search for Army Intelligence," *Isis* 84, no. 2 (1993): 278–309.

27. Terman, *Measurement of Intelligence*, 6–7.

28. Robert M. Yerkes, "Eugenic Bearing of Measurements of Intelligence," *Eugenics Review* 24, no. 4 (1923): 225–45.

29. Harry Hamilton Laughlin, *Eugenical Sterilization in the United States* (Chicago: Psychopathic Laboratory of the Municipal Court of Chicago, 1922); Kenneth M. Ludmerer, "Genetics, Eugenics, and the Immigration Restriction Act of 1924," *Bulletin of the History of Medicine* 46, no. 1 (1972): 59–81; P. K. Wilson, "Harry Laughlin's Eugenic Crusade to Control the 'Socially Inadequate' in Progressive Era America," *Patterns of Prejudice* 36, no. 1 (2022): 49–67.

30. These were Alabama (1919), Arizona (1929), California (1909), Connecticut (1909), Delaware (1923), Georgia (1937), Idaho (1925), Indiana (1907), Iowa (1911), Kansas (1913), Maine (1925), Michigan

(1913), Minnesota (1925), Mississippi (1928), Montana (1923), Nebraska (1915), New Hampshire (1917), New York (1912; declared unconstitutional in 1918), North Carolina (1919), North Dakota (1913), Oklahoma (1931), Oregon (1917), South Carolina (1935), South Dakota (1917), Utah (1925), Vermont (1931), Virginia (1924), Washington (1909), West Virginia (1929), and Wisconsin (1913).

31. Human Betterment Foundation, "Human Sterilization Today," 1940, folder "Human Betterment Foundation," box 1, Bronson Price Papers, American Philosophical Society, Philadelphia. For sterilization after World War II, see Rebecca M. Kluchin, *Fit to Be Tied: Sterilization and Reproductive Rights in America, 1950–1980* (New Brunswick, NJ: Rutgers University Press, 2011).

32. See, for example, Alleyne Ireland, *Democracy and the Human Equation* (New York: E. P. Dutton, 1921); William McDougall, *Is America Safe for Democracy?* (New York: Charles Scribner's Sons, 1921); Lothrop Stoddard, *The Revolt against Civilization: The Menace of the Under Man* (New York: Charles Scribner's Sons, 1922).

33. Nicholas Pastore, "The Army Intelligence Tests and Walter Lippmann," *Journal of the History of the Behavioral Sciences* 14, no. 4 (1978): 316–27.

34. Walter Lippmann, "The Mental Age of Americans," *New Republic* 32, no. 412 (1922): 213–15; Lippmann, "The Mystery of the 'A' Men," *New Republic* 32, no. 413 (1922): 246–48; Lippmann, "The Reliability of Intelligence Tests," *New Republic* 32, no. 414 (1922): 275–77; Lippmann, "The Abuse of the Tests," *New Republic* 32, no. 415 (1922): 297–98; Lippmann, "Tests of Hereditary Intelligence," *New Republic* 32, no 416 (1922): 328–30; Lippmann, "A Future for the Tests," *New Republic* 32, no. 417 (1922): 9–11.

35. Lewis Madison Terman, *Genetic Studies of Genius*, vol. 1, *Mental and Physical Traits of a Thousand Gifted Children* (Stanford, CA: Stanford University Press, 1926).

36. Joel N. Shurkin, *Terman's Kids: The Groundbreaking Study of How the Gifted Grow Up* (Boston: Little, Brown, 1992).

37. Ronald A. Fisher, "The Correlation between Relatives on the Supposition of Mendelian Inheritance," *Transactions of the Royal Society of Edinburgh* 52 (1918): 399–433; Fisher, "The Causes of Human Variability," *Eugenics Review* 10, no. 4 (1919): 213–20.

38. Jay Laurence Lush, *Animal Breeding Plans* (Ames, IA: Collegiate Press, 1943). For a critique of this measure, see Richard C. Lewontin, "Annotation: The Analysis of Variance and the Analysis of Causes," *American Journal of Human Genetics* 26 (1974): 400–411.

39. Sewall Wright, "Correlation and Causation," *Journal of Agricultural Research* 20, no. 7 (1921): 557–85; Barbara Burks, "The Relative Influence of Nature and Nurture upon Mental Development: A Comparative Study of Foster Parent-Foster Child Resemblance and True Parent-True Child Resemblance," *Twenty-Seventh Yearbook of the National Society for the Study of Education*, part 1 (1928): 219–316; Karl J. Holzinger, "The Relative Effect of Nature and Nurture Influences on Twin Differences," *Journal of Educational Psychology* 20, no. 4 (1929): 241–48.

40. Lewis Madison Terman, "Suggestions for the Preparation of Material for the 1928 Yearbook on the Possibilities and Limitations of Training," folder "Terman, Lewis Madison," box 89, Charles B. Davenport Papers.

41. Lewis Madison Terman, "The Influence of Nature and Nurture upon Intelligence Scores: An Evaluation of the Evidence in Part I of the 1928 Yearbook of the National Society for the Study of Education," *Journal of Educational Psychology* 19, no. 6 (1928): 362–73, 370.

42. Carl Campbell Brigham, "Intelligence Tests of Immigrant Groups," *Psychological Review* 37, no. 3 (1930): 158–65, 165. Goddard had come to a similar conclusion about a decade earlier. Intelligence tests he performed at Ellis Island showed that about 80 percent of incoming immigrants were feebleminded, but follow-up studies showed them to be adapting well to their new surroundings. Zenderland, *Measuring Minds*.

43. Terman, *Measurement of Intelligence*.

44. Otto Klineberg, *Negro Intelligence and Selective Migration* (New York: Columbia University Press, 1935); Klineberg, *Race Differences* (New York: Harper and Brothers, 1935).

45. Elazar Barkan, *The Retreat of Scientific Racism: Changing Concepts of Race in Britain and the United States between the World Wars* (New York: Cambridge University Press, 1992).

46. For more on the elder Osborn, see Jonathan Spiro, *Defending the Master Race: Conservation, Eugenics, and the Legacy of Madison Grant* (Lebanon, NH: University Press of New England, 2009).

47. Frederick Henry Osborn, "Notes for a Eugenic Program," November 14, 1934, folder "Frederick Osborn—Papers #4," box 17, American Eugenics Society Records, American Philosophical Society, Philadelphia.

48. James Angell, quoted in AES annual meeting program, 1932, box 1, American Eugenics Society Records.

49. Frederick Henry Osborn, "Application of Measures of Quality," 1936, folder "Frederick Osborn—Papers #7," box 17, American Eugenics Society Records; Osborn, "Measures of Quality in the Study of Population," 1936, folder "Frederick Osborn—Papers #6," box 17, American Eugenics Society Records.

50. Frederick Henry Osborn, "Development of a Eugenic Philosophy," *American Sociological Review* 2, no. 3 (1937): 389–97, 389.

51. Frederick Henry Osborn, "Draft and Comments re Eugenic Program," April 10, 1935, folder "Frederick Osborn—Papers #5," box 17, American Eugenics Society Records.

52. Frederick Henry Osborn, "A Eugenics Program for the United States," 1935, folder "Frederick Osborn—Papers #3," box 17, American Eugenics Society Records.

53. Osborn, "Draft and Comments."

54. Osborn, "Draft and Comments."

55. Frederick Henry Osborn, "Moral Responsibilities of Parenthood," May 20, 1936, folder "Frederick Osborn—Papers #6," box 17, American Eugenics Society Records.

56. Emily Klancher Merchant, *Building the Population Bomb* (New York: Oxford University Press, 2021); Nathaniel Comfort, *The Science of Perfection: How Genes Became the Heart of American Medicine* (New Haven, CT: Yale University Press, 2012); Alexandra Minna Stern, *Telling Genes: The Story of Genetic Counseling in America* (Baltimore: Johns Hopkins University Press, 2012); Richard H. Osborne and Barbara T. Osborne, "The Founding of the Behavior Genetics Association, 1966–1971," *Social Biology* 46, no. 3–4 (1999): 207–18.

57. Frederick Henry Osborn, letter to Alexander Robertson, folder "Milbank Memorial Fund Grants, 1963–1967," box 18, American Eugenics Society Records.

58. American Eugenics Society, Princeton conference transcripts, boxes 8–11, American Eugenics Society Records.

59. American Eugenics Society, "Five Year Report," 1971, American Eugenics Society Records.

60. Louise Erlenmeyer-Kimling and Lissy F. Jarvik, "Genetics and Intelligence: A Review," *Science* 142, no. 3598 (December 13, 1963): 1477–79.

61. Howard F. Taylor, *The IQ Game: A Methodological Inquiry into the Heredity-Environment Controversy* (New Brunswick, NJ: Rutgers University Press, 1980).

62. Arthur R. Jensen, "Estimation of the Limits of Heritability of Traits by Comparison of Monozygotic and Dizygotic Twins," *Proceedings of the National Academy of Sciences* 58, no. 1 (1967): 149–56.

63. Jensen, "Estimation of the Limits of Heritability," 151–53, emphasis in the original.

64. R. M. Cooper and John P. Zubek, "Effects of Enriched and Restricted Early Environments on the Learning Ability of Bright and Dull Rats," *Canadian Journal of Psychology* 12, no. 3 (1958): 159–64.

65. Jensen, "Estimation of the Limits of Heritability," 153.

66. Arthur R. Jensen, "How Much Can We Boost IQ and Scholastic Achievement?," *Harvard Educational Review* 39, no. 1 (1969): 1–124.

67. Joel N. Shurkin, *Broken Genius: The Rise and Fall of William Shockley, Creator of the Electronic Age* (New York: Palgrave Macmillan, 2006).

68. William H. Tucker, *The Funding of Scientific Racism: Wickliffe Draper and the Pioneer Fund* (Champaign: University of Illinois Press, 2002).

69. Walter F. Bodmer and Luigi Luca Cavalli-Sforza, "Intelligence and Race," *Scientific American*, October 1, 1970, https://scientificamerican.com/article/intelligence-and-race.

70. Richard C. Lewontin, "Race and Intelligence," *Bulletin of the Atomic Scientists* 26, no. 3 (1970): 2–8, 7.

71. Ned Block, "How Heritability Misleads about Race," *Cognition* 56 (1995): 99–128.

72. Lewontin, "Annotation."

73. Leon Kamin, *The Science and Politics of IQ* (New York: Halsted Press, 1974).

74. Taylor, *IQ Game.*

75. Douglas S. Falconer, *Introduction to Quantitative Genetics* (New York: Ronald Press, 1960), 185.

76. Jensen, "Estimation of the Limits of Heritability," 151.

77. Christopher Jencks, Marshall Smith, Henry Acland, Mary Jo Bane, David Cohen, Herbert Gintis, Barbara Heyns, and Stephan Michelson, *Inequality: A Reassessment of the Effect of Family and Schooling in America* (New York: Basic Books, 1972), 66–67.

78. For examples of today's behavior geneticists arguing that Jencks's example represents a valid genetic cause, see Kathryn Paige Harden, *The Genetic Lottery: Why DNA Matters for Social Equality* (Princeton, NJ: Princeton University Press, 2021).

79. Sandra Scarr, "On Arthur Jensen's Integrity," *Intelligence* 26, no. 3 (1998): 227–32.

80. United Nations, "Convention on the Prevention and Punishment of the Crime of Genocide" (1948).

81. Peter M. Blau and Otis Dudley Duncan, *The American Occupational Structure* (Hoboken, NJ: John Wiley and Sons, 1967); Robert M. Hauser, "Educational Stratification in the United States," *Sociological Inquiry* 40, no. 2 (1970): 102–29.

82. Aaron Panofsky, *Misbehaving Science: Controversy and the Development of Behavior Genetics* (Chicago: University of Chicago Press, 2014).

83. Eric Turkheimer and Irving I. Gottesman, "Is h^2 = 0 a Null Hypothesis Anymore?," *Behavioral and Brain Sciences* 14, no. 3 (1991): 410–11.

84. David C. Rowe, *The Limits of Family Influence: Genes, Experience, and Behavior* (New York: Guilford Press, 1994).

85. Robert Plomin and C. S. Bergeman, "The Nature of Nurture: Genetic Influence on 'Environmental' Measures," *Behavioral and Brain Sciences* 14, no. 3 (1991): 414–27.

86. Today's behavior geneticists continue to make misleading claims about the implications of heritability. See, for example, Kathryn Asbury and Robert Plomin, *G Is for Gene: The Impact of Genetics on Education and Achievement* (London: John Wiley and Sons, 2013).

87. John C. Loehlin, letter to Elizabeth S. Russell, April 22, 1975, folder "Correspondence—Loehlin, John C.," box 37, Genetics Society of America Archives, American Philosophical Society, Philadelphia.

88. Elizabeth S. Russell, "Statement of GSA Members on Heredity, Race, and IQ," January 26, 1976, folder "Committee on Genetics, Race, and Intelligence—Draft Resolutions #2," box 37, Genetics Society of America Archives.

89. Panofsky, *Misbehaving Science.*

90. *Population and Environment* no longer publishes explicitly racist research. For that story, see Emily Klancher Merchant, "Environmental Malthusianism and Demography," *Social Studies of Science* 52, no. 4 (2022): 536–60.

91. Richard Herrnstein and Charles Murray, *The Bell Curve: Intelligence and Class Structure in American Life* (New York: Free Press, 1994).

92. Linda S. Gottfredson, "Mainstream Science on Intelligence: An Editorial with 52 Signatories, History, and Bibliography," *Intelligence* 24, no. 1 (1994): 13–23.

93. *Science News* staff, "Some Past *Science News* Coverage Was Racist and Sexist. We're Deeply Sorry," *Science News*, March 24, 2022, https://www.sciencenews.org/article/past-science-news-coverage-racism-sexism.

94. Only women married to men could qualify. David Plotz, *The Genius Factory: The Curious History of the Nobel Prize Sperm Bank* (New York: Penguin Random House, 2005).

95. No children were conceived from the sperm of any of the three Nobel Prize winners who donated to the repository. Plotz, *Genius Factory*.

96. For more on intelligence, eugenics, and sperm donation, see Cynthia R. Daniels and Janet Golden, "Procreative Compounds: Popular Eugenics, Artificial Insemination and the Rise of the American Sperm Banking Industry," *Journal of Social History* 38, no. 1 (2004): 5–27.

97. American Eugenics Party, correspondence with the American Eugenics Society, 1964–1968, box 1, American Eugenics Society Records.

98. American Eugenics Society, meeting minutes, 1972, box 7, American Eugenics Society Records. The organization is now called the Society for Biodemography and Social Biology.

99. Notable exceptions include Kevles, *In the Name of Eugenics*; Stern, *Eugenic Nation*; and Ordover, *American Eugenics*.

100. Aaron Panofsky's *Misbehaving Science* downplays the eugenic origins of behavior genetics, likely because it is based on interviews with behavior geneticists in the early 2000s, after this disavowal was complete. For examples of behavior geneticists citing Galton as the founder of their field, see Jeffrey W. Gilger, "Contributions and Promise of Human Behavioral Genetics," *Human Biology* 72, no. 1 (2000): 229–55, and Robert Plomin and Richard Rende, "Human Behavioral Genetics," *Annual Review of Psychology* 42, no. 1 (1991): 161–90.

101. Aaron Panofksy, "From Behavior Genetics to Postgenomics," in *Postgenomics: Perspectives on Biology after the Genome*, ed. Sarah Richardson and Hallam Stevens (Durham, NC: Duke University Press, 2015), 150–73.

102. Jason D. Boardman and Jason M. Fletcher, "Evaluating the Continued Integration of Genetics into Medical Sociology," *Journal of Health and Social Behavior* 62, no. 3 (2021): 401–81.

103. For the engagement of demography with genetics, see Emily Klancher Merchant, "Of DNA and Demography," in *Recent Trends in Demographic Data*, ed. Parfait M. Eloundou-Enyegue (London: IntechOpen, 2023), https://intchopen.com/online-first/1129796.

104. Dalton Conley and Jason M. Fletcher, *The Genome Factor: What the Social Genomics Revolution Reveals about Ourselves, Our History, and the Future* (Princeton, NJ: Princeton University Press, 2017); Jeremy Freese, "The Arrival of Social Science Genomics," *Contemporary Sociology* 47, no. 5 (2018): 524–36.

105. Avshalom Caspi, Joseph McClay, Terrie E. Moffitt, Jonathan Mill, Judy Martin, Ian W. Craig, Alan Taylor, and Richie Poulton, "Role of Genotype in the Cycle of Violence in Maltreated Children," *Science* 297 (2002): 851–54.

106. Brett C. Haberstick, Jeffrey M. Lessem, John K. Hewitt, Andrew Smolen, Christian J. Hopfer, Carolyn T. Halpern, Ley A. Killeya-Jones, Jason D. Boardman, Joyce Tabor, Ilene Siegler, Redford B. Williams, and Kathleen Mullan Harris, "MAOA Genotype, Childhood Maltreatment, and Their Interaction in the Etiology of Adult Antisocial Behaviors," *Biological Psychiatry* 75, no. 1 (2014): 25–30.

107. Christopher F. Chabris, B. M. Hebert, Daniel J. Benjamin, Jonathan P. Beauchamp, David Cesarini, M. van der Loos, M. Johannesson, P. K. Magnusson, P. Lichtestein, C. S. Atwood, Jeremy Freese, T. S. Hauser, Robert M. Hauser, Nicholas Christakis, and David I. Laibson, "Most Reported Genetic Associations with General Intelligence Are Probably False Positives," *Psychological Science* 23, no. 11 (2012): 1314–23.

108. Christopher F. Chabris, James J. Lee, Daniel J. Benjamin, Jonathan P. Beauchamp, Edward L. Glaeser, Gregoire Borst, Steven Pinker, and David I. Laibson, "Why It Is Hard to Find Genes Associated with Social Science Traits: Theoretical and Empirical Considerations," *American Journal of Public Health* 103, Supplement 1 (2013): S152–66.

109. Christopher F. Chabris, James J. Lee, David Cesarini, Daniel J. Benjamin, and David I. Laibson, "The Fourth Law of Behavior Genetics," *Current Directions in Psychological Science* 24, no. 4 (2015): 304–12.

110. Cornelius A. Rietveld, Sarah E. Medland, Jaime Derringer, Jian Yang, Tõnu Esko, Nicolas W. Martin, Harm-Jan Westra, Konstantin Shakhabazov, Abdel Abdellaoui, Arpana Agrawal, et al., "GWAS of 126,559 Individuals Identifies Genetic Variants Associated with Educational Attainment," *Science* 340, no. 6139 (May 30, 2013): 1467–71.

111. James Devitt, "Genes Have Small Effect on Length of Education," *Futurity*, June 3, 2013, https://www.futurity.org/genes-have-small-effect-on-length-of-education.

112. Paul Voosen, "There Is No Gene for Finishing College," *Chronicle of Higher Education*, May 30, 2013, https://www.chronicle.com/blogs/percolator/there-is-no-gene-for-finishing-college.

113. Robert M. Sapolsky, "A Height Gene? One for Smarts? Don't Bet on It," *Wall Street Journal*, January 31, 2014, https://ssgac.org/documents/New-Evidence-That-Genes-Play-a-Minimal-Role-in-Many-Traits-WSJ.pdf.

114. Aysu Okbay, Jonathan P. Beauchamp, Mark Alan Fontana, James J. Lee, Tune H. Pers, Cornelius A. Rietveld, Patrick Turley, Guo-Bo Chen, Valur Emilsson, S. Fleur W. Meddens, et al., "Genome-Wide Association Study Identifies 74 Loci Associated with Educational Attainment," *Nature* 533 (2016): 539–42; James J. Lee, Robbee Wedow, Aysu Okbay, Edward Kong, Omeed Maghzian, Meghan Zacher, Tuan Anh Nguyen-Viet, Peter Bowers, Julia Sidorenko, Richard Karlsson Linnér, et al., "Gene Discovery and Polygenic Prediction from a 1.1-Million-Person-GWAS of Educational Attainment," *Nature Genetics* 50, no. 8 (August 2018): 1112–21.

115. Conley and Fletcher, *Genome Factor*; Plomin, *Blueprint*; Harden, *Genetic Lottery*.

116. Aysu Okbay, Yeda Wu, Nancy Wang, Hariharan Jayashankar, Michael Bennett, Seyed Moeen Nehzati, Julia Sidorenko, Hyeokmoon Kweon, Grant Goldman, Tamara Gjorgjieva, et al., "Polygenic Prediction of Educational Attainment within and between Families from Genome-Wide Association Analyses in 3 Million Individuals," *Nature Genetics* 54 (2022): 437–49.

117. Okbay et al., "Polygenic Prediction of Educational Attainment." This estimate was validated by Laurence J. Howe, Michel G. Nivard, Tim T. Morris, Ailin F. Hansen, Humaira Rasheed, Yoonsu Cho, Geetha Chittoor, Rafael Ahlskog, Penelope A. Lind, Teemu Palvainen, Matthijs D. van der Zee, Rosa Cheesman, et al., "Within-Sibship Genome-Wide Association Analyses Decrease Bias in Estimates of Direct Genetic Effects," *Nature Genetics* 54 (2022): 581–92.

118. Behavior geneticists still estimate the heritability of intelligence to be about 0.8, but since educational attainment is a more complex social outcome, its heritability is lower.

119. Michelle N. Meyer, Paul S. Appelbaum, Daniel J. Benjamin, Shawneequa L. Callier, Nathaniel Comfort, Dalton Conley, Jeremy Freese, Nanibaa' A. Garrison, Evelynn M. Hammonds, K. Paige Harden, Sandra Soo-Jin Lee, Alicia R. Martin, Daphne Oluwaseun Martschenko, Benjamin M. Neale, Rohan H. C. Palmer, James Tabery, Eric Turkheimer, Patrick Turley, and Erik Parens, "Wrestling with Social and Behavioral Genomics: Risks, Potential Benefits, and Ethical Responsibility," *Hastings Center Report* 53, no. S1 (March–April 2023): S2–49.

120. Nicholas W. Papageorge and Kevin Thom, "Genes, Education, and Labor Market Outcomes: Evidence from the Health and Retirement Study," *National Bureau of Economic Research Working Paper* (2018), https://nber.org/papers/w25114.

121. Kathryn Paige Harden, Benjamin W. Domingue, Daniel W. Belsky, Jason D. Boardman, Robert Crosnoe, Margherita Malanchini, Michel Nivard, Elliot M. Tucker-Drob, and Kathleen Mullan-Harris, "Genetic Associations with Mathematics Tracking and Persistence in Secondary School," *Science of Learning* 5, no. 1 (2020) 1, https://nature.com/articles/s41539-020-0060-2.

122. Brenna Henn, Emily Klancher Merchant, Anne O'Connor, and Tina Rulli, "Why DNA Is No Key to Social Equality: On Kathryn Paige Harden's *The Genetic Lottery*," *LA Review of Books*, September 21, 2021, https://lareviewofbooks.org/article/why-dna-is-no-key-to-social-equality-on-kathryn-paige-hardens-the-genetic-lottery.

123. Alice B. Popejoy and Stephanie M. Fullerton, "Genomics Is Failing on Diversity," *Nature* 538, no. 7624 (2016): 161–64.

124. Alicia R. Martin, Christopher R. Gignoux, Raymond K. Walters, Genevieve L. Wojcik, Benjamin M. Neale, Simon Gravel, Mark J. Daly, Carlos D. Bustamante, and Eimear E. Kenney, "Human Demographic History Impacts Genetic Risk Prediction across Diverse Populations," *American Journal of Human Genetics* 100, no. 40 (2017): 635–49.

125. See, for example, Harden, *Genetic Lottery*.

126. Jeremy Freese, Ben Domingue, Sam Trejo, Kamil Sicinski, and Pamela Herd, "Problems with a Causal Interpretation of Polygenic Score Differences between Jewish and Non-Jewish Respondents in the Wisconsin Longitudinal Study," SocArXiv (2019), https://osf.io/preprints/socarxiv/eh9tq; Makhamanesh Mostafavi, Arbel Harpak, Ipsita Agarwal, Dalton Conley, Jonathan K. Pritchard, and Molly Przeworski, "Variable Prediction Accuracy of Polygenic Scores within an Ancestry Group," *eLife* 9 (2020): e48376.

127. Two examples of this misrepresentation are Plomin, *Blueprint*, and Harden, *Genetic Lottery*. For a critique, see Erik Turkheimer, "Peter Visscher on the Genomics of Complex Human Traits," May 12, 2022, http://turkheimer.com/peter-visscher-on-the-genomics-of-complex-human-traits.

128. This is particularly true in Plomin, *Blueprint*.

129. Graham Coop and Molly Przeworski explain how Paige Harden does this in their review of *The Genetic Lottery*. Coop and Przeworski, "Lottery, Luck, or Legacy: A Review of *The Genetic Lottery: Why DNA Matters for Social Equality*," *Evolution* 76, no. 4 (2022): 846–53.

130. Plomin, *Blueprint*.

131. Dalton Conley, "What If Tinder Showed Your IQ?," *Nautilus*, September 10, 2015, https://nautil.us/what-if-tinder-showed-your-iq-235618.

132. This is just a reminder that no company is currently offering polygenic embryo screening for educational attainment. Parents who want to screen embryos on this basis would need to obtain their raw data from one of these companies and then upload it to a separate website to calculate their embryos' polygenic scores for educational attainment.

133. For a recent assertion of these ideas, see Gregory Clark, "The Inheritance of Social Status: England 1600 to 2022," *Proceedings of the National Academy of Sciences* 120, no. 27 (June 26, 2023): e2300926120; James J. Lee, "The Heritability and Persistence of Social Class in England," *Proceedings of the National Academy of Sciences* 120, no. 29 (July 5, 2023): e2309250120.

134. Turley et al., "Problems with Using Polygenic Scores."

135. See, for example, Jordan Lasker, Bryan J. Pesta, John G. R. Fuerst, and Emil O. W. Kirkegaard, "Global Ancestry and Cognitive Ability," *Psych* 1, no. 1 (2019): 431–59.

136. For an example of this kind of racist garbage, see J. Juerst, V. Shibaev, and E. O. W. Kirkegaard, "A Genetic Hypothesis for American Race/Ethnic Differences in Mean g: A Reply to Warne (2021) with Fifteen New Empirical Tests Using the ABCD Dataset," *Mankind Quarterly* 63, no. 4 (June 2023): 527–600.

137. Robbee Wedow, Daphne O. Martschenko, and Sam Trejo, "Scientists Must Consider the Risk of Racist Misappropriation of Research," *Scientific American*, May 26, 2022, https://scientificamerican.com/article/scientists-must-consider-the-risk-of-racist-misappropriation-of-research; Emily Klancher Merchant, "Hold Science to Higher Standards on Racism," *STAT*, June 30, 2022, https://statnews.com/2022/06/20/hold-science-to-higher-standards-on-racism.

Race and Reproduction

6

Evangelical Christianity, Race,
and Reproduction

Meaghan O'Keefe

Many people have called racism "America's original sin." In Christianity, original
sin is the doctrine that all human beings are tainted from birth with a tendency
toward sin. It's worth paying attention to the reference here—racism is present
from the inception of America, and we are innately driven toward it. This inclina-
tion toward injustice is cast in the religious language of sin. The theological underpinnings of what we now think of as racial categories are undeniable, as is the
connection between race and chattel slavery. Indeed, scholars generally agree that
"race was a product rather than the cause of American slaveholding."[1] As noted in
the chapter by Lisa Ikemoto, race-based slavery gave rise to particular ideas about
moral and religious capacity based upon physical characteristics that became rei-
fied in race science. The formation of these racial categories, however, was not
a steady and clear path away from religion[2] and toward a secular and scientific
notion of biological race.[3] Rather, ideas about race over the last four centuries
(and, indeed, before that as well) are part of a complex set of ongoing interactions
that result in sometimes fragmented, sometimes congruent, and more often contingent and inconsistent ideas of what race is and what race does.

The concept of race is itself both a "product" of its social context and "pro-
ductive," in the sense that it continues to organize personal experiences, scientific
knowledge, and political action.[4] In practice, the United States is a profoundly
racialized country, meaning that people are always, in one way or another, assigned
a racial identity and that identity structures the relationships, opportunities,
and experiences available to people.[5] My argument centers on white Evangelicals
for two reasons. First, when it comes to issues of race, white Evangelicals have
very different beliefs and experiences from non-white Evangelicals; and, second,

they have a great deal of political power to enforce those beliefs.[6] For example, Republican political positions have become nearly indistinguishable from Evangelical belief. It is not just that white Evangelicals are overwhelmingly Republican: white Evangelicalism's ideas about race and racism, gender norms, anti-statism, and insistence on the goodness of America have proved to be a major draw for people whose politics align with these beliefs. To illustrate, a recent study showed that people who voted for Trump and were not Evangelical in 2016 were more likely to identify as Evangelical in 2020.[7] In this case, as in others, political motivations are inseparable from social, intellectual, and religious ones. Put simply, understanding the complex ways religion affects and is affected by social and political goals is crucial to gaining insight into how a large subset of Americans make sense of DNA, race, and reproduction and how they make and defend political choices about these issues.

Many assume that white Evangelicals are opposed to the science of genetics and hold conservative views on racial equity. This oppositional take is partially right but misses much of the nuance. Some recent studies illustrate the complicated ways contemporary white American Evangelicals think about race, genetics, and the biological sciences. For example, although white Evangelicals are generally opposed to evolution, they do accept that genetic tests reveal where a person's ancestors may have lived.[8] Some conservative Evangelicals—while accepting direct-to-consumer genetic testing as legitimate—argue that we are all descended from Adam and Eve and even go so far as to use Punnett squares (a diagram that predicts genotypes in breeding experiments)[9] to present human genomes as evidence of the truth of biblical narratives.[10] These sources argue that all possible human genetic diversity was present in Adam and Eve (a belief that the geneticist Joseph L. Graves Jr. has described as "scientifically impossible"[11]) and that phenotypic differences can be traced to the biblical dispersions of people. Recent surveys also reveal interesting juxtapositions of acceptance of certain kinds of scientific expertise but not others. For example, Evangelical Christians are less likely to be suspicious of genetically modified foods than members of other faith traditions[12] but more likely to believe that scientists are overstating harms when it comes to climate change.[13] When it comes to gender and sex, white Evangelicals are the group most likely to believe that gender is set by the sex assigned at birth,[14] but, interestingly, a large proportion (46 percent) of those who believe sex at birth determines gender say they learned this from "science."[15] This is in line with popular white Evangelical views that emphasize the importance of biological or chromosomal sex as part of the theological idea of complementarity—that God created men and women for different but complementary responsibilities and roles and that this is reflected not just in social expectations but in bodies themselves.

White Evangelicals are committed to the idea that biological sex is fixed and absolutely essential to a virtuous life and a moral world. Race is understood as primarily biological, but, in contrast to sex, it is a source of division, not the basis for a moral order. Prejudice against people because of their skin color is considered a sin. Additionally, Evangelicals believe that seeing people as different because they belong to different racial groups undermines the idea that we are all made in the image of God. In this worldview, racism is mainly an individual problem, not an institutional or a systemic one. Recognizing race is also suspect in that it draws attention to divisions between people rather than seeing all people as the children of God. Thus, in many Evangelical communities, racism is talked about as a "sin problem, not a skin problem." This formulation does three things: it makes racism ancillary to the problem of sin; it reduces race to a merely phenotypic difference; and it frames racism as an individual moral failing rather than a systemic problem.[16] Treating race as a merely cosmetic—"skin-deep"—difference minimizes the harms done and the power encoded in such classifications. This decoupling of racism from larger forces allows it to be transported to the realm of individual sin.[17] The remedy, then, is for individual people to recognize and repent for their wrongdoing, not advocate for systemic change. This erasure of racialized systems has the added effect of not simply dismissing the experience of racial inequality but actually assigning blame to people who suffer in racist systems. For example, white Evangelicals tend to blame poverty on Black people choosing not to value marriage and raising children,[18] rather than seeing the breakup of families as the result of mass incarceration and the foster care system's systematic targeting of Black families.[19] As I show later in this chapter, for white Evangelicals, these beliefs are justified through a mode of understanding based on interpreting scripture's relationship to the material world and a particular theology of sin and responsibility. They are also inextricably linked to ideas about race.

The history of race in the United States is inseparable from the history of American Christianity, particularly Protestant sects. Protestant Christianity dominated the colonies and the early republic. This continued into the nineteenth century: the 1860 US Census found that 95 percent of places of worship in the United States were Protestant. Within Protestantism, Evangelical Christianity is and has been a particularly strong force in the development of ideas about race and racism in the United States. Historically, American Evangelicalism emerged as a dominant sect in the early nineteenth century, following the First and Second Great Awakenings.[20] These two religious movements were characterized by emotionalism; direct, personal engagement with the Bible; and a strong emphasis on the supernatural; and believers were deeply engaged in personal, spiritual transformation.[21] These practices and beliefs are still central to white Evangelicalism, and they form much of conservative political thinking on the topic of race and reproduction.

Looking at any one of these topics in isolation without considering the theological substrate and historical contexts would result in partial and seemingly incoherent positions. Contemporary Evangelical Christian ideas about race are not, however, evidence of an attachment to unwavering historical precedent. Neither are they strictly contemporary. This attachment to the past while engaging with scientific research and contemporary political issues is part of the tendency of religions to repeat and reconfigure traditions and practices while maintaining a semblance of constancy. In other words, for many faith communities, religion is understood to be unchanging yet always present and engaged. As Kathryn Lofton observes, religious people, institutions, and communities are engaged in "reiteration and repetition (and, yes, revision)" as they connect past practices, texts, and beliefs to their "lived religious present."[22] Essentially, contemporary white Evangelical Christians, like many other religious communities, are engaged in a constant process of adapting, accommodating, or rejecting systems of knowledge as they apply their beliefs and traditions to current issues. While much of the analysis in this chapter engages with theological arguments, such arguments never exist outside of or prior to social and political contexts: "religion is part and parcel of racial, ethnic, class, and gender inequality."[23] Theological arguments are themselves tools and products of racialization.

RELIGIOUS AND RACIAL CATEGORIES

In 2010 Franklin Graham—the son of the famous Evangelical preacher Billy Graham—said of Barack Obama, "I think the president's problem is that he was a Muslim, his father was a Muslim. The seed of Islam is passed through the father like the seed of Judaism is passed through the mother."[24] In actuality, Obama is a Christian, but it is not accidental that the United States' first black president was identified as a religious "other": studies have shown the racist undertones and motivations for characterizing Obama in this way.[25] Just as racial and religious othering here are not new, the relationship of religious inheritance to racial categorization also has deep connections to the past.

The notion of a kind of "hereditary heathenism" helps explain how non-white people were initially relegated to the fixed and heritable category of heathen and, in the mid-seventeenth century, how it also "invented an entirely new concept—what it meant to be 'white.'"[26] In this period, racial categories had not been cemented in the way they are now, but religious categories were well established. Early colonists used the categories of heathen and Christian to mark differences and enforce legal separation between the English colonists and Native Americans and enslaved Africans. These conditions were understood to be heritable, with one preacher remarking that the children of such heathens were neither "baptizable nor pardoned" and therefore could not claim the privileges afforded white colonists even if they were to convert.[27] This declaration was, in part, a reaction to the practice of freeing enslaved people who converted to Christianity.[28] This "loophole" was legally done away with by the Virginia colonists in 1667 when they declared that baptism did not automatically confer freedom for enslaved people.[29] The weakening of the association between being a Christian and being white, however, necessitated new legal categories. For example, a 1705 Virginia law forbade the "whipping of a 'christian white servant naked.'"[30] In this example, it was not sufficient to identify someone as Christian to signify white.[31] In the same year, Virginia colonists also declared that "negroes, mulattoes, and Indian servants" could not serve as witnesses in court. Previous versions of this law simply declared that non-Christians (this category included Catholics) were forbidden from testifying; the new version kept the explicit prohibition on Catholics but found it necessary to add racial classifications as well. The passing of this law was part of the process through which colonists enshrined racial ideology.[32] More generally in this period, religious categories formed the basis for racial ones, and other colonies, such as Puritan New England and the West Indies, showed a similar tendency to merge religious categories with race-based ones.[33]

These race-based categories were not shorn of their religious significance. Quite the opposite: racial classification was always caught up with religious concerns and epistemologies, and the older conceptions did not disappear—they were simply reconfigured. For example, the category of heathen was still used for much of the nineteenth and twentieth centuries as a means of differentiating white people from non-white people. It remains useful for "racial clumping," meaning the grouping of culturally different people together in the category of non-white, a practice that served as means of asserting the spiritual and racial superiority of white Protestants in the contemporary United States.[34]

RACE AND SLAVERY

The creation of racialized religious legal categories was certainly not the only religious means of constructing and maintaining whiteness. Biblical explanations of the dispersion of peoples accompanied these legal categorizations and, as mentioned above, still feature prominently in the contemporary Evangelical narratives explaining human difference. These differences are attributed variously to people being the descendants of Cain (one of the sons of Adam and Eve) or of the sons of Noah, or God's destruction of the Tower of Babel. Each of these explanations comes with negative connotations. Cain killed his brother, Abel, and attempted to lie to God about it. God then cursed Cain and condemned him to wander as a fugitive and "marked" him.[35] This "mark" was interpreted by some to be dark skin. Ham, a son of Noah, saw his father drunk and naked and did not cover him as his brothers, Shem and Japheth, did.[36] Noah cursed Ham, saying that for his transgression, Ham's sons would be the slaves of his brothers.[37] Following this passage, there is a genealogy of Noah's sons, describing their dispersal and the civilizations they founded. Historically and in the present, many Christians have explained human diversity through the different lineages of Noah's sons. For example, some argued that Europeans are the descendants of Japheth, Asians are the descendants of Shem, and Africans are from the lineage of Ham.[38] Other biblical stories were also used to explain differences among human groups. Both historically and in contemporary Evangelical discourse, differences among people are traced to the destruction of the Tower of Babel.[39] In this story, the peoples of the world all speak the same language, and they come together to build a tower to the "heavens" that they might rival the power of God. God sees this and causes their language to become different and unintelligible so that they can no longer cooperate in building the tower, and he then disperses the people throughout the world.[40] Historically, Evangelical Christians used the stories of Cain and Ham to justify enslavement and the Tower of Babel to support segregation. Present-day accounts tend to downplay the idea that the descendants of Ham or Cain[41] carry a hereditary curse, but the wrongdoing of these figures and the notion of generational inheritance of the physical marks of sin are never far away, especially for those familiar with biblical texts.

While the stories of Cain, Ham, and the Tower of Babel presuppose that all human beings are descended from Adam and Eve (a theory known as monogenism), another explanation circulated during this period: polygenism, the idea that different human groups had different origins, also described in the chapter by Lisa Ikemoto. This theory, although seemingly at odds with biblical accounts of human origins, was deeply rooted in religious belief and biblical interpretation. Indeed, the first comprehensive account of polygenism was written in Latin by Isaac de La Peyrère in 1655. This work was, in part, an attempt to explain, if Adam and Eve had only sons, whom Cain married and had children with.[42] La Peyrère's explanation was that there must have been people already in existence before the creation of Adam and Eve. His work was condemned by both Catholic and Protestant authorities as heretical, but it enjoyed broad popularity, with four reprints issued and Dutch and English translations.[43] Over the next two centuries, La Peyrère's ideas were more attacked than supported.[44] In the colonies of Virginia and Barbados, however, slaveholders used this argument to religiously justify enslavement on the grounds that people of African descent were not truly human and therefore incapable of becoming Christian.[45]

In the nineteenth-century United States, these ideas were resurrected and combined with scientific ideas about racial difference.[46] American polygenists used measurements of physical characteristics to reinforce existing ideas about the different "races."[47] Using these observations, Samuel Morton, a Philadelphia physician who authored a central text on the topic, concluded that it was "highly unlikely"[48] that human beings shared a common ancestor. Morton's ideas were taken up and expanded upon by Josiah Nott, one of the preeminent physicians of the nineteenth-century American South.[49] Nott drew on Morton's empirical observations to conclude that non-Europeans were biologically inferior. He developed this argument further, reasoning that, given these differences, it was undeniable that Europeans were the only descendants of Noah. Nott rejected the idea that people of African descent were the sons of Ham, and, although he seems to have been silent on the question of Cain, he explicitly rejected the idea of a "universal

Adam" that all human beings descended from.[50] Samuel Cartwright, a southern physician in favor of a scientific approach to the question of race, took the project of reconciling polygenism with scientific racism further, using the story of Eve and the forbidden fruit to argue for a separate creation of different groups of people. In the book of Genesis, Adam and Eve live in paradise (the Garden of Eden), and they are free to do whatever they like (there is no sin in this world) except eat the fruit of the tree of knowledge. A serpent persuades Eve to go against God's wishes and eat the fruit. She not only eats the fruit, she persuades Adam to eat it as well, and God punishes them by casting them out of the Garden of Eden. This "Fall" is understood as the moment when death, suffering, and sin enter the world. According to Cartwright, the tempter in Eden was not a serpent but, as he puts it, a "negro gardener" who, unlike the other beasts, has the ability to speak.[51] Cartwright attached scientific ideas about racial difference to this supposed second race, which he described as human but "more like the monkey" than other kinds of humans.[52]

Ideas like Nott's and Cartwright's grew in popularity during the 1850s as a means of justifying enslavement. There was, however, significant pushback by other pro-slavery Southern Christians, who dismissed polygenism as not only heretical but also potentially damaging to the institution of slavery.[53] For these proslavery figures, monogenism was entirely compatible with enslavement and even mandated by it. In their version of monogenism, all human beings were descended from Adam and Eve, but God had different plans for different peoples. In this view, the story of Ham is not simply an explanation of why some people were enslaved, but also carries the implication that to enslave people is to enact God's plan.[54] They also looked to the Hebrew Bible patriarchs who had extended households that included enslaved people[55] for a religiously sanctioned model of slavery.[56] They argued for slavery as a "divine institution" instantiated in an "ideal of the master-slave relationship," which held that the paternalistic regard slaveholders had for enslaved people was morally superior to the impersonal "wage slavery" of the North.[57] Not only was the institution of slavery held up as part of God's plan, it was individually good for enslaved people because it allowed Christianity to save their souls.[58]

EMANCIPATION AND SEGREGATION

Once slavery was abolished, there was great concern about how a society with free Black people would function. In the Reconstruction era and after, extrajudicial actions such as lynching were part of a broad campaign of terror designed to keep Black people from claiming their rights. On the legal side, while initially restrained by federal control during Reconstruction, Southerners soon enacted Jim Crow laws that legally segregated Black and white Americans. These legal and illegal efforts were designed to maintain white power and, as Lisa Ikemoto's chapter has discussed, white purity. Fears about interracial sex and marriage were common tropes. The "Black Codes" that were enacted after the end of Reconstruction made intermarriage illegal in all Southern states, and public justifications for lynching often featured accusations of sexual violence by Black men against white women. These ideas were justified religiously. For example, in 1867 Buckner Payne, a Southern clergyman, took up Cartwright's assertion of a separate pre-Adamite race.[59] Payne argued that this other race was complicit in much of the other wrongdoing described in scripture, such as the construction of the Tower of Babel.[60] Moreover, Payne claimed that the near total destruction of humanity by flood was God's punishment for miscegenation. In the story of the flood, God sees that the earth is "corrupt" and filled with "violence" and decides that he will destroy the earth and all the people except for Noah and his family, who are instructed to build an ark so that they may survive the flood.[61] According to Payne, what God really objected to were the offspring of Adam and the other created race, and these were the people he chose to destroy. Payne also posits that the ark contained the pure white individuals of Noah's family as well as members of the "black race," whom Payne claims were also on the ark but as "beasts" rather than persons. Thus, Payne argued, allowing marriage between white and Black people invited biblical retribution. Over the next 30 years, Payne's arguments were repeated, adapted, and added to in order to maintain the idea of a separate creation for white and non-white people and to justify segregation and, more particularly, to condemn miscegenation.[62]

Objections to miscegenation and the belief that white people were physically different from and superior to non-white people were not limited to proponents of polygenism. In the latter part of the nineteenth century, Southern Christians embraced different versions of a kind of monogenic polygenism. Prior to emancipation, one of the more prevalent proslavery arguments explained that the sons of Ham (meaning people of African descent) had fallen into a kind of physical and moral degeneracy and were thus in need of the "benign stewardship" of chattel slavery.[63] In this version, physical differences were interpreted as heritable deformities that had arisen after the descendants of Noah had settled in various parts of the world. Thus, all human beings are descended from Adam and Eve, but some groups of people have undergone biological changes along with moral degradation. These explanations were not abandoned with slavery; they were adapted to argue against miscegenation.

In the late nineteenth century, Protestants in the South did not just rely on scripture for justifications; they also drew on the race science of the day to defend their positions. The idea that people of African descent had undergone some process of degeneration resurfaced with the new science of eugenics—described in the chapters by Mark Fedyk, Lisa Ikemoto, and Emily Klancher Merchant—used as the primary justification. One instructive example of this is the Southern Baptist adoption of race science. The Southern Baptist Convention is not just the largest

Evangelical denomination; it is the largest Protestant denomination in the United States and has one of the more fraught and visible histories with slavery and racial discrimination. It was explicitly founded on the basis of support for slavery, and the arguments marshalled by the Southern Baptists in favor of enslavement and, later, segregation are an amalgam of biblical, Evangelical (in the sense of conversion), and scientific ones.

Proslavery Baptists had argued variously that slavery was part of God's plan to bring Christianity to Africans, that it was blessed by the apostle Paul,[64] that people of African descent carried the "mark of Cain," or that they were the children of Ham. They soundly rejected polygenism as heretical. For Southern Baptists, and Southern Protestants more generally, religious arguments in favor of the nature of and role for people of African descent predominated.[65] By the 1890s, however, the faculty of the Southern Baptist Seminary firmly grounded their arguments about black inferiority in race science.[66] Some, such as John Broadus, the second president of the Southern Baptist Seminary, argued for a kind of Lamarckian inheritance of moral capacity, intelligence, and industry, arguing that the parts of Africa from which most enslaved people were captured lacked all civilization and that centuries of barbarianism had cemented negative characteristics that were both biological and heritable.[67] Others, such as Charles Gardner, a professor of sociology and homiletics, were more specific, stating that intellectual and moral capacities were transmitted "physiologically." Gardner went on, explaining that manifestations of social progress in people of African descent were the result of "receiving the blood of higher races into [their] veins" and that by a process of "natural selection . . . in proportion as the negro race ceases to be negro we may expect its capacity for progress proportionately to increase."[68] In this quote, we see one side of miscegenation: white men fathering children with Black women "improved" the Black "race." What goes unsaid is the corollary, that white women having children with Black men would degrade the white "race." While the racism baked into these statements should be familiar, the broad embrace of the principles of eugenics—the idea that some human beings are superior to others, and the superior ones ought to be encouraged to breed more and the inferior types should be discouraged from breeding—may seem surprising to contemporary readers who might associate white Evangelicals with hostility to ideas of evolution. These religious positions, however, were not unusual in this period.

In actuality, what is perhaps most surprising about religion and eugenics in the United States is not that there was religious opposition to it but rather the remarkable alacrity with which the Protestant establishment embraced the eugenics movement, almost from its inception. As Christine Rosen shows in her history of the role American religious leaders played in the eugenics movement, not only did Protestant clergy support the means and ends of eugenics, the scientific eugenics movement also enthusiastically engaged with clergy.[69] The American Eugenics Society (AES) had a Committee on Cooperation with Clergymen whose members included many of the most prominent religious figures of the era.[70] Starting in 1926, the AES sponsored a sermon contest that awarded $500 for first prize, $200 for second, and $100 for third, significant amounts in the 1920s.[71] Part of the motivation for recruiting religious leaders was their influence on public discourse, but another major goal was to encourage people of "the better classes"[72] to reproduce more. These people were referred to as "builders," and eugenicists had determined that a large proportion of church members were in this category.[73]

While the majority of the clergy involved with the AES were recruited from mainline Protestant churches, it's important to note that support for eugenics was not limited to Protestant clergy: some Reform Jewish rabbis also supported the eugenics movement. Catholic officials were also involved. John A. Ryan, a Catholic priest and the onetime head of the National Catholic Welfare Conference's Social Action Department, served on the AES's Committee on Cooperation with Clergymen alongside another Catholic priest, John M. Cooper.[74] These Catholic members, however, were more reticent about the means used to accomplish eugenic goals, and they resigned in 1930 after Pius XI issued a formal condemnation of sterilization in his encyclical on marriage, *Casti Connubii*.[75]

Many Protestant proponents framed their support for eugenics as part of the "Social Gospel," a belief that it was the duty of humanity to prepare the world for the second coming of Christ. This meant that active social reform—including "health" interventions such as eugenics—was the means through which the salvation of the world could be achieved.[76] Other, more conservative, Protestants came to see social reform as a dangerous diversion from the spiritual mandate to save souls and even as a heresy in that it imagined that human effort could transform the world.[77] Evangelical Protestants initially accepted eugenics as an explanation of the social hierarchies and a means of social progress, but their interest in eugenics soon dissipated, and between 1900 and 1930, many conservative Evangelicals, particularly Fundamentalists, Pentecostals, and fervent Evangelicals, came to outright reject eugenics.[78]

Just as the issue of slavery split the Baptist Church into the American Baptist Church in the North and the Southern Baptist Church, these responses were shaped as much or more by geography and social context as by theology. For example, Northern Baptists were much more supportive of the efforts of eugenicists and expressed deep concern about "race suicide," while Southern Baptists were less enthusiastic.[79] Southern Baptists believed in a hierarchy of races and the value of "Anglo-Saxon" stock but rejected efforts to regulate marriage for eugenic purposes. This was in part because there was less immigration in the South and because segregation was quite effective at separating races already.[80] Another factor in Southern resistance to eugenics was that eugenicists identified the embrace of "primitive religion" as a marker of Anglo-Saxon degeneration. By "primitive religion," they meant Pentecostal practices such as speaking in tongues and Evangelical "ecstatic religious revivals," both of which were very popular among

Southern whites.[81] Finally, rural whites, especially poor ones, were often the targets of eugenicist efforts to "improve" the Anglo-Saxon race, and, therefore, poor religious white people were deeply suspicious of such efforts.

INTEGRATION

When it came to desegregation, a similar amalgam of religious argument and race science resurfaces. After the *Brown v. Board of Education* Supreme Court ruling, which declared segregation in public schools unconstitutional, there was a backlash by Southern Christians.[82] Prior to the ruling, there had been some tentative support for desegregation among Southern Methodists and Southern Baptists, but when desegregation became reality, Southern Evangelicals rejected racial moderation.[83] Initial arguments within the denominations centered on scriptural and religious arguments. They cited verses such as Deuteronomy 7:3–4, which cautions the Israelites not to intermarry with neighboring groups. One Southern Baptist leader expressed his fears that desegregation would lead to interracial marriage, declaring, "Negroes are descendants of Ham [and] we whites must keep our blood pure."[84] More general arguments that God had created different races and separated them by continents also featured prominently. A Baptist church in South Carolina issued this statement: "God meant for people of different races to maintain their race purity and racial indentity [*sic*] . . . God has determined the 'bounds of their habitation' [Acts 17:26]."[85] Others brought up the "mark" of Cain and the Tower of Babel as justifications. None of these ideas are new; indeed, we have seen them from the very beginning of the American colonies.

These religious arguments share a common fear of miscegenation. When Little Rock High School was forcibly integrated in 1957, much of the anti-integration rhetoric centered on sexual threats to white girls.[86] This fear permeated discussions of desegregation and were viewed by many, even outside the South, as reasonable. For example, President Eisenhower, who later sent federal troops to Little Rock High School to enforce desegregation, reportedly said to Chief Justice Earl Warren while *Brown v. Board of Education* was being decided, "these [people opposed to desegregation] are not bad people. All they are concerned about is to see that their sweet little girls are not required to sit in school alongside some big overgrown Negroes."[87] While many of the arguments were presented as rooted in long tradition and the deep scriptural precedent, the fears that motivated them were very much in the present.

These religious arguments were part of primarily a moral justification for fellow believers. When it came to legal challenges to *Brown v. Board of Education* and the 1964 Civil Rights Act's ban on racial discrimination in public accommodation, other arguments were brought to bear. They included race science claims about brain differences between African Americans and European Americans[88] (which are now discredited) and claims that public accommodation laws for restaurants violated the Constitution's freedom of religion cause.[89] Arguments against public accommodation and public school integration that relied on religious freedom and pseudoscientific racism were dismissed by the Supreme Court. In response to the failure of these legal efforts, Southern segregationists employed a number of strategies designed to defund public schools and transfer those resources to private, white-only schools, known as "segregation academies."[90] These schools were not all officially religiously affiliated, but, by their own admission, "religion [was] an integral part of the [private] school movement."[91] In 1976 such schools were declared in violation of civil rights statutes,[92] and in 1983 the Supreme Court ruled that religious schools that had segregationist policies were no longer entitled to tax-exempt status.[93] In spite of this loss, some schools still maintained segregationist policies. Indeed, it wasn't until 2000 that Bob Jones University, a private Evangelical university in South Carolina and the main petitioner in the 1983 case, revoked its ban on interracial dating. While these attitudes and positions have significant staying power, the pseudoscientific and biblically justified racist arguments fell out of favor. As a result, in the decades following the 1964 Civil Rights Act, public arguments against integration became more covert.

In the period following the civil rights movement, many white Southern Baptists as well as other Evangelicals and mainline Protestants adopted a message that love modeled on Jesus's love for all humanity would solve the problem of racism.[94] This message called for individuals to change their hearts and be open to loving all people equally. The emphasis on interior, sincere change carried with it opposition to change forced from the outside. Love must be freely given and freely chosen. Loving all human beings equally meant seeing each of them as a child of God rather than a member of a particular race. The emphasis on interior transformation centered redemption as the overriding message. This focus on the "ritual of self-redemption"[95] made the experience of white people repenting in private the defining one in the problem of racism. There is an additional dimension, however, that further displaces the experiences of people of color. For white Evangelicals (and others as well), racism is prejudice and, as such, can be experienced by anyone who feels they have been treated badly because of their skin color. In the words of a white Pentecostal woman, racism isn't just "whites against blacks. Blacks do not like white people."[96] In this quote, racism is about people not liking one another. Placing racism in the realm of the personal also, not infrequently, included framing it as a mutual problem that could be solved through interpersonal interactions.[97]

The rapid shift in the Southern Baptist Convention from an avowed prosegregation position to this seemingly radical acceptance would be remarkable if it had, in fact, desegregated congregations. In practice, most Southern churches—and American churches more generally—remained and remain highly segregated. This resistance to integration within churches was, in many ways, built into the model of individual spiritual transformation. First, because interior transformation is, by its nature, a solitary activity, interacting with people of different racial backgrounds is somewhat ancillary. It may change the nature of those interactions, but it is not the starting point of change. There is also something of a sleight of hand working in the view that, because we are all children of God, our individual differences are a hindrance to achieving the love that Jesus intended. This view means that mentioning race becomes a way of reinforcing difference and undermining love for all. In other words, if only we stopped talking about race, racism could be eliminated.

What this stance translates into politically is that if, for instance, a Black person were to say, "this is an issue that affects Black people as a group," then the implication is that the person is actually reinforcing racism because they are insisting that people's group identity is somehow more important than their individual worth. In other words, pointing out the general experience of Black people is a way of only seeing people's color instead of treating each person as an individual, worthy of respect. Making recognition of the social reality of racism the problem expands the possibilities for pursuing political goals that disproportionately harm minoritized people. For example, in a 1981 interview, Lee Atwater, one of the most prominent Republican strategists of the 1980s, explained how racially coded tactics worked, saying that in 1954, you could just say "N—, N—, N—" and in 1964, you had to switch to saying "forced busing, states' rights" and in the 1980s, you had to talk about economic policies in which "blacks get hurt worse than whites" but as long as you didn't mention race, these policies could be glossed as color-blind.[98] Thus, arguments in favor of family values, such as an emphasis on parental rights in education—which in the 1970s meant the right to attend private, segregation academies[99]—could be framed as simply moral choices rather than actions inseparable from racial politics.

CONTEMPORARY IDEAS ON RACE AND RACISM

Contemporary white Evangelical discourse about race and human diversity contains some now familiar topics—the creation of Adam and Eve, the Tower of Babel, Noah and his sons—but they have been grafted onto a more expansive, racialized, and biologized theological stance. This stance allows for a more thoroughly worked-out theology that makes talking about race the source of racism rather than a means of addressing it. It also integrates old ideas of human difference with genetic science into a system of belief that centers the patriarchal nuclear family as the site and source of moral action. The use of "family values" as a rallying cry allows for coded racism, but it also expands political possibilities by using the "inviolability" of the family as a means of reinforcing gender norms and shaping reproductive policies. In what follows, I trace these ideas through contemporary texts by prominent conservative white Evangelicals.

In 2001 John MacArthur, the pastor of a Southern California megachurch and the host of *Grace to You*, a national Evangelical television and radio program, preached a sermon titled "The Sins of Noah." The sermon brings up polygenism in order to condemn it but goes on to resurrect the monogenist idea of degeneration of the lineage of Ham, Noah's son. MacArthur states that those who claim that there were humanlike creatures before the creation of Adam and Eve are unequivocally wrong: we have all descended from Adam through Noah and his sons. He argues that when "evolutionists" tell us that some groups of humans diverged from other populations between 40,000 and 60,000 years ago,[100] what they are saying is that aboriginal Australians and Native Americans are "spiritless, soulless hominids" because Adam was the first creature with a soul and he (as well as the earth) was created 6,168 years ago.[101] This amalgamation of science and scripture continues with MacArthur arguing that the differences between people come from "culture and adaption." What he means by this is that, when people were dispersed after the destruction of the Tower of Babel, small groups became isolated, and, over time, certain genetic features became fixed in these populations. He claims that some of these changes were adaptive, such as the change from darker to lighter skin in order to absorb vitamin D, and some were the result of genetic drift, but all went according to God's plan. As MacArthur puts it, God "sorted the gene pool out exactly the way that He wanted to sort it." What MacArthur describes here is a morally neutral process of adaptation and change. When it comes to culture, however, MacArthur has a version of the older idea of the degeneracy of Ham.

MacArthur attributes the degeneracy of all humanity to our propensity to sin. It is sin that causes defective genes and sin that causes human degradation. MacArthur is clear that all human beings are sinners and, to some degree or other, degraded. All of his examples, however, are of non-European people. He lists "pygmies in Africa," "Hottentots in South Africa," people from Papua New Guinea, and "aboriginal people in Australia" as examples of "degeneration" and characterizes them as people "so far gone" that it is nearly impossible to "preach the Gospel" to them. This degeneracy threatens to overtake Western civilization, which was once becoming better but now is headed in the wrong direction, and soon, he predicts, "we're going to be stark naked, running around with a spear, stabbing people." The message here is clear: although all human sin and the state of current civilizations are the result of God's plan, the cultures of darker-skinned people have degraded further and their degradation threatens European ascendancy. MacArthur brings in new scientific insights—and, not incidentally, accuses evolutionary scientists of considering some people as less than human—to explain human difference, but the overall story is not a new one.

Although this sermon is still available in both video and text on the *Grace to You* website, this kind of transparently racist discourse is far less common in current conservative white Evangelical discourse. Mainstream contemporary white Evangelicals condemn the idea that the story of Ham means that white people are racially superior. John Piper, an influential conservative white Evangelical minister,[102] refutes this straw man argument, not by addressing the question of subservience or enslavement but, instead, by explaining that Ham is not the father of Africans. Other are more straightforward in their condemnation of the "curse of Ham." Albert Mohler, the president of the Southern Baptist Theological Seminary, goes further, declaring that interpreting the "curse of Ham" as a biblical endorsement of racial superiority "reflects such ignorance of Scripture and such shameful exegesis" that it constitutes heresy, meaning that it is in opposition to essential Christian beliefs.

These leaders may reject the older, explicitly white supremacist arguments based on the "curse of Ham," but they still attribute human difference to degeneracy and sin, albeit in somewhat less offensive ways than MacArthur's sermon. In addressing a church member's question about whether the Tower of Babel was the beginning of racial differentiation, Piper takes the question quite literally and unequivocally states that the rebellion of the builders of the Tower of Babel was the "immediate cause" of the geographic and linguistic diversity we see in the world.[103] According to Piper, human differentiation is punishment. There is, however, more to this story. Piper goes on to say that this differentiation was part of God's plan for redemption; the "evil in the world" that results in different ethnicities is present so that Jesus could bring them together. In this version, Christians must proselytize to all nations to bring them together in faith. In other words, conversion erases the differences between human beings. United in love of God, people of all backgrounds become "brothers and sisters."[104]

Nearly all conservative white Evangelical leaders attribute human diversity to the effects of human sin in the Garden of Eden, the "wickedness" and "violence"[105] that prompted God to destroy all humanity except for Noah and his family, and the defiance in building the Tower of Babel. What is interesting is that, like MacArthur above, they also accept that these differences are genetic. Leaders like Mohler accept the science behind identifying genetic mutations, observing that "in every individual human genome, there are genetic errors."[106] The proximate cause of these mutations may be attributed to biological processes, but the ultimate cause is humanity's fall in the Garden of Eden. As Mohler explains, "in Eden, in the perfection of creation, there would have been nothing wrong with a single human genome."[107] Once sin came into the world, all was corrupted, and when mutations occur in the human genome, it is because "human genetic structure . . . [is] affected by and corrupted by sin."[108] For conservative Evangelicals, the "fact of sin" is not only the cause of the material structure of genomes, it is also the reason for human behavior. Scientists are in error when they try to attribute human behavior to "genes and chromosomes" because, according to Billy Graham Ministries, they have "fail[ed] to give a proper place to the inborn twist toward selfishness, viciousness, and indifference to God, making many of their conclusions only pseudoscientific."[109] Interestingly, in this passage, scientific facts are made to fit a theological conception, but they are also used as a reason to invalidate science on its own terms—science that doesn't consider sin is "pseudoscientific."

Part of the concern over attributing sinful behavior to genetics is because of the "born this way" argument used by many LGBTQ activists. Many conservative Evangelicals deny that there is any genetic basis for nonheterosexual orientations.[110] Others are agnostic on the question but reiterate that, even if it were genetic, that doesn't matter; people must still abstain from sinful behavior. A common trope in this line of reasoning is the same one just mentioned: genetic disease and mutation are a result of the Fall, and simply because something is found in nature, it does not constitute a moral explanation because creation itself has been "corrupted and distorted by sin."[111] More generally, these leaders accept the mechanics of genetics but refuse to imbue it with any specific moral significance. In this scenario, LGBTQ sexuality and gender identity are signs of fallenness, whether they have origins in biological difference or not. Race, however, is treated as unquestionably biological and primarily about skin color. The biology of race is the result of God's punishment, either in the dispersal of people after the destruction of the Tower of Babel or simply a result of being cast out of the Garden of Eden.

The biblical explanations of human diversity and genetic mutation we see here are deeply rooted in the concept of the Fall and human sin. Human differences in terms of sex and gender, however, are categorically different. Conservative Evangelicals' discussions about the proper social roles for men and women begin with Genesis 1:27: "And God created man in his own image, in the image of God created he him; male and female he created them."[112] This verse is the foundation for believing in separate and different roles for men and women. The key theological difference here is that this differentiation occurred before Adam and Eve disobeyed God and were expelled from the Garden of Eden. The garden was a world and a place without sin, and creation before the Fall was exactly as God intended it. Thus, deviation from both social roles or rigidly defined biological sex is sinful.

When it comes to differences in sex, conservative Evangelicals wholeheartedly embrace biological determinism. While much of the discussion centers on social roles with men as leaders and women as nurturers, much recent discussion has centered on physical differences in male and female bodies, especially in light of the recent opposition to gender-affirming care and transgender rights. For example, in a recent guest post on John Piper's site, Stephen Wedgeworth argues that "biblical manhood and womanhood" is inscribed on a molecular and cellular level and that God-given sexual differences are manifest on a genetic and hormonal level.[113] The Nashville Statement—an Evangelical declaration on gender roles, sex, and marriage and their sanctity and signed by Mohler, Piper, and MacArthur—is even more biologically focused, stating that "the differences between male and female reproductive structures are integral to God's design for self-conception as male or female."[114] The differences between men and women exist in order to fulfill God's command to "be fruitful and multiply": as Mohler explains, "reproductive success and the obedience to the reproductive command that God gives us, depends upon men being men and women being women."[115] In this system of understanding, sexual dimorphism exists because God has ordained it, and biblical accounts are given as the reason scientific evidence is the way it is.

The apotheosis of this differentiation is "biblical" marriage and reproduction. It is also in the Evangelical theology of marriage that we see the theological justification for treating effects of racial discrimination as the fault of the victims of it. Evangelical ideas about marriage are tied to particular ideas about how faith and morality are formed. White Evangelical biblical marriage is based on the theological concept of "relationality." Relationality is the central tenet of American white Evangelicalism and holds that salvation can come only through a personal relationship with Jesus.[116] This divine/human relationship is then "transposed"[117] onto interpersonal relationships as "love and respect [for Jesus] overflows into our love and respect for our neighbors."[118] Marriage holds a special place in this configuration. It is a reflection of divine wholeness[119] and is patterned on love for Jesus, and it is the source of moral decision-making. It follows, then, that immoral decisions are made by people in the wrong sort of relationships (i.e., those not shaped by love of Jesus and structured by opposite-sex marriage). This results in a worldview in which problems such as poverty stem from failed relationships, and the solution is thought to be personal rather than systemic change. The "family values" embedded in the idea of biblical marriage are inseparable from conservative historical opposition to the civil rights movement and to contemporary positions on police violence and racism.[120]

Focusing on personal relationships while avoiding discussion of structural racism is consonant with the underlying structure of what the sociologist Eduardo Bonilla-Silva has called "racism without racists," a set of explanations and justifications created by white Americans that resolve "the apparent contradiction between [white people's] professed colorblindness and the United States' color-coded inequality."[121] Scholars of American Evangelicalism have noted this phenomenon in the resurgence of racism under the cover of sexual morality, remarking that, while white supremacy in its older forms has been delegitimized in much of public discourse, religious ideas about gender and sexual morality have been "grafted" onto "patriotic[122] and racial traditionalism."[123] To put it another way, with the demise of de jure segregation, biblical marriage (with the man as the head) provides the organizing principle for a properly ordered traditional society that reinforces de facto racial inequality.

From a theological perspective, conservative Evangelical Christian discourse has generally treated racism as an issue of personal responsibility and individual decisions and actions.[124] In this view, there are political consequences to ideas about racism, but it is essentially a moral and theological issue. This creates a set of circumstances where conservative white Evangelical leaders can condemn racism without recognizing the institutional and systemic processes by which racism is maintained. For example, Albert Mohler, the president of the Southern Baptist Theological Seminary, declares that "white superiority . . . .is a heresy."[125] Rick

Warren, the head pastor of Saddleback Church, which has a weekly attendance of 23,000, says that racism is "a sin problem, not a skin problem."[126] Franklin Graham calls racism "an evil" and states that, to God, "no skin color is more or less important."[127] In each of these examples, the speakers focus on skin color: race is framed as a cosmetic difference.

While these Evangelical leaders explicitly address racism, others mention racism but as the starting point for other arguments rather than the main focus. For all of these men, racism is treating people badly because of their skin color; it is not the cause of racial inequality. Like Warren and Mohler, John MacArthur calls racism a sin but quickly pivots to discussing masculinity and fatherhood, stating that the lack of Black fathers present in their children's lives "is a holocaust" and the only "hope for peace in society is masculine, virtuous men."[128] Similarly, a guest article on Piper's website uses the topic of violence against Black people as an opportunity to talk about abortion, stating that "it's illegal to murder George Floyd, but it's legal to murder preborn George Floyds. And it happens over 800,000 times a year in the United States."[129] In these examples, racism is the lesser sin. The real sins are abortion and men's failure to "act like men."[130]

One theme that emerges in these texts is that the topic of racism serves as a kind of jumping-off point for discussion of other issues the speakers deem important. Part of this may be a symptom of white people's general discomfort around the topic of race,[131] but there are also particular kinds of theological framing that makes this kind of switch in topics both coherent and logical within a particular understanding of sin. Pivoting from racism to the absence of Black fathers makes a certain kind of sense if one sees biblical marriage as essential to right behavior. Such a framing locates the problem in personal (and sinful) decisions about marriage and fatherhood rather than systemic structures that damage and undermine Black families. Other topics, such as abortion, are also tied into ideas of marriage and reproduction. Abortion perverts the essential role of women as child bearers. More generally, abortion and racism are both sins, and sins—by their very nature—are the actions of individuals, so it is individuals alone who will have to answer before God for their transgressions.[132] White Evangelicals' emphasis on individual sin, the paramount importance of the right kind of relationships, the understanding of humanity as irredeemably fallen, and belief that only the saving grace of God can remedy the ills of prejudice all figure in white Evangelicals' stance on systemic racism in the United States.

For all of the leaders studied, racism is a sin, and each of them calls on Christians to repent and to love one another. Sin is understood as ever-present and human beings as essentially and primarily depraved. As Driscoll puts it, "sin is not just we do, but who we are."[133] Sin permeates human society. Keller explains that when human beings turn away from God, they make idols of other things like race or culture, which results in inequity and injustice.[134] Along similar lines, Piper states that individual sin always results in "systemic or structural" sin and that all human institutions are "permeated with sin" and "reflect, embody, preserve, and advance" sin.[135] Albert Mohler also argues in this same vein, observing that "sin corrupts every single human system in one way or another, because it's made up of sinful human beings." Mohler explains that while it is individual sinners who seek out or perform abortions, "human society" is "made up of those sinners influenced by those sinners, legislated by those sinners, bring[ing] the sin into the structures and systems of society."[136] He goes on to say that racism is also present in human institutions for the same reasons. Unlike many of the other preachers discussed here, Mohler talks about "systemic racism" but says that if we are to think about structural sin, then we ought to "start with something like the scandal of abortion, the horror of the legal murder of the unborn."[137] He argues that "radical abortion rights legislation" has systematically transformed American culture into a "culture of death." Mohler argues that since abortion has corrupted an entire society, it is hardly surprising that other sins might also affect social institutions.

It is clear that some of these men believe that racism is structural, systemic, and deeply sinful. In their view, however, it is just one kind of sin among many. For example, MacArthur lists "sexual immorality, relentless assault of feminism, overexposure to perversion, complete collapse of homes" as both the cause and the result of evil "abound[ing] absolutely everywhere."[138] Treating racism as merely one manifestation of the overwhelming presence of sin and evil allows speakers to quickly pivot to other sins that they see as endemic to a sinful society, such as abortion or the absence of biological fathers in children's lives.

This classic Protestant pessimism about the depravity of the world works against politically and institutionally oriented solutions to the problem of systemic inequality. Albert Mohler explains most clearly why political and social responses are not only ineffective but also "dangerous."[139] The problem is that imagining "improvement is possible in human society" replaces the transcendent good news of salvation with an earthly ambition.[140] Believing in this kind of improvement, Mohler contends, is buying into the "fundamentally false belief" that we can eliminate sin from society.[141] In this view, true change must begin with those united with Christ by faith.[142] All of these men call on people to treat those different from themselves with love and respect. These actions, however, must be preceded by repenting to God for your sins, and then, transformed by God's grace, a person may begin to change the world around them.[143] Thus, change is grounded in salvation and evangelization. As Franklin Graham puts it, "we are to tell a hurting world that Jesus shed His blood and died for our sins" and that in turning to Christ we will be saved and filled "with the love that conquers racism and hatred."[144] The emphasis on individual sin also works to preclude the possibility of calling out racism in individuals and institutions.

By emphasizing personal, interior transformation as part of a relationship between an individual and God, there is very little space for criticism from other people. Many of the speakers actively discourage pointing out racist behavior in other people. For example, Keller criticizes those who call attention to racial injustice, saying that they "resort to shaming and often exhibit a self-righteous manner," an approach he criticizes as unbiblical.[145] Driscoll goes further, criticizing those who "continually march for justice, demand wrongs be made right, and argue ad nauseum [*sic*] on social media about systemic sin using all the various -isms (racism, sexism, nationalism, classism, ageism, etc.)" as hypocrites.[146] Not only are people discouraged from criticizing others in order to avoid hypocrisy, there is also an underlying belief that to point out differences—including racial discrimination—is counter to Christ's vision for the Church as the unity of all believers.[147] Warren goes further, saying that the Church is a family; you are "called to belong to the Church," and to think of yourself as a "visitor" or a "stranger" is to place yourself in opposition to God's will.[148] This means that bringing up racism means that you are actively resisting God's plan for unity. This emphasis on the unity of the Church, combined with the discouragement of calling attention to racist actions, works against reform within churches.

To summarize, in these texts, race itself is often presented as simply skin color, racism reduced to discriminating against someone explicitly due to the color of their skin, and racial inequality is due to poor individual decisions. Racism is condemned but often used as a means to pivot to other issues such as abortion. Claims of racism within the Evangelical churches are often viewed as disloyal and destructive to God's intent for a unified people of God. In a larger sense, Evangelical Christians' belief in the paramount importance of personal relationships as well as "accountable freewill individualism" (which holds that people are "individually in control of, and responsible for, their own destinies")[149] leads them to discount larger social forces, such as lack of access to education, employment discrimination, and racial profiling by law enforcement. Thus, white Evangelicals are far more likely to attribute economic disparities between whites and Blacks to poor personal choices.[150] When it comes to government intervention to address racial inequality—or indeed many sorts of social problems—most white Evangelicals see such programs as "naive, wasteful, misguided, sinful, and often counteracting real solutions."[151] Poor personal choices are understood to be rooted in the wrong kind of relationships, so government programs are more likely to actually compound rather than solve these problems.

CONCLUSION: THE TRIUMPH OF FAMILIAL LOVE

Their unwillingness to address problems of racial inequality and injustice does not mean that white Evangelicals do not take action about what they see as the problem of racial division. As mentioned above, preachers call upon their members to cast the sin of racism from their hearts and reach across racial lines to "demonstrate the power of biblical unity."[152] Inclusion is important to white Evangelical churches, but the framing of inclusion relies on tropes of sameness, such as we are all the same under the skin.[153] Most Evangelical churches, however, remain deeply segregated, and although the number of Hispanic members in majority-white Evangelical churches has increased, the number of Black members has remained extremely low.[154] Church leadership has remained segregated, with white people significantly overrepresented as pastors.[155] People of color in majority-white Evangelical spaces often experience racial microaggressions but are discouraged from bringing up issues of racial discrimination and mistreatment. It would seem from these conditions that the project of white Evangelical racial reconciliation has failed. If the goal, however, is to make white people feel as if they are doing something about racism as they understand it, it has succeeded.

One particular practice stands out as unusual in its triumphalism about racial reconciliation: transracial adoption. Transracial adoption moves the experience of race even farther from the realm of institutions and systems. It places it within the family. As a white preacher says about his Black and white sons, "racism isn't a social issue. It's a family issue."[156] He goes on to say that this is "what racial reconciliation and familial love that transcends skin color looks like."[157] Transracial adoption has been held up as a means to "grow God's family" and to achieve racial harmony through bringing a non-white child into a white Christian family formed by a biblical marriage.[158] Beginning in the early 2000s, the white Evangelical community has become more and more interested in transracial adoption as an imperative of faith. Transracial adoption is presented as a means of rescuing "orphans" and a way of furthering racial harmony. It does not, however, involve the integration of differing cultural systems—after all, babies don't have culture—but is rather the wholesale subsuming of a non-white child into white Evangelical culture.[159] Adoption, as Perry and Whitehead pointedly observe, involves "social and legal uniting of racial groups in a situation where one has guardianship over the other"[160] rather than a relationship between equals. Thus, the difference between the adopted child and the adopting family is phenotypic rather than social or cultural.

This same rescue and reconciliation narrative has extended into embryo adoption (when couples "adopt" embryos created during IVF but not used by the couple who created them), which shows more starkly the reduction of race to phenotype. Within this movement, there is a more recent trend among white Evangelicals to request non-white embryos as a means of addressing "racial conflict."[161] In both cases, transracial adoption and the use of donated embryos, there is an underlying belief that racial harmony can be achieved by sidestepping the lived experience of race within racialized cultural systems. What is clear here is that race is understood as genetic and biological, not socially constructed. Children are emblems of difference but acultural ones. Thus, racial harmony is achieved by attaching significance to phenotype without the presence of culture. Race is reproduced as a phenotypic difference within the enveloping world of whiteness: the brokenness of a fallen world is healed by a white biblical family's love.

For contemporary white Evangelicals, family, marriage, and personal responsibility have become the main framing device for talking about social issues. Such framing allows racism to become less visible to white Christians, and, as Jemar Tisby very generously puts it, this framing "often leads them [white Evangelicals] to unknowingly compromise with racism."[162] More pointedly, endorsing policies that are not explicitly racist but that disproportionately hurt people of color, white conservatives can "proclaim their racial innocence."[163] The commitment to the fantasy of racial innocence takes different forms. Indeed, some have suggested that, contrary to the popular narrative that abortion is what brought white Evangelicals into the political sphere as a cohesive voting bloc, it was the stripping of tax-exempt status from racially discriminatory religious schools (such as Bob Jones University) that actually motivated white Evangelicals.[164] Scholars contend that opposition to abortion and the championing of "family values" has not replaced racism, it has simply camouflaged it.[165]

While there is no doubt that many conservative white Evangelicals are sincerely opposed to abortion and would be distressed to be accused of racism, many scholars of American Evangelicalism have long contended that issues of race have structured American religious belief.[166] Some argue that white Christian churches in the United States have been the driving force in maintaining "white supremacy and resist[ing] black equality"[167] and that this is particularly true of white Evangelicalism.[168] While many white Evangelicals would object to being accused of maintaining white supremacy, there is evidence that they view racial discrimination and racialized violence differently than most Americans. The majority of Americans believe that police officers treat Black people differently than white people, but the majority of Evangelicals believe the opposite: that police officers treat Black people no better or no worse than white people.[169] They are also more likely to view police shootings of unarmed Black people as "isolated incidents" rather than as part of a broader pattern.[170] These views have proved remarkably stable, with little change from 2015 to 2020.[171] Scholars of Evangelicalism have expressed profound pessimism about the ability of white Evangelicals to meaningfully engage with the problems of racism in United States. They give this bleak assessment, "white evangelicalism does more to perpetuate the racialized society than to reduce it," because the very structures by which white Evangelicals understand the world make them both incapable of recognizing the existence of systemic racism and unable to take action against it.[172]

Michael S. Hamilton argues that white Evangelical social concerns are "disconnected from their theology."[173] His explanation is that, because white, Black, and Hispanic Evangelicals share a belief in the centrality of Christ's sacrifice, the centrality of scripture, the need for conversion, and activism to promulgate these beliefs, it follows then that the political beliefs of white Evangelicals cannot be theological. Hamilton is right if one adopts a very narrow definition of Evangelicalism and if one thinks of it as aspirational rather than descriptive. In other words, if we imagine the religious beliefs as something to live up to, to strive for, then, as Hamilton sees it, supporting Donald Trump is difficult to square with the public piety and personal morality that many Evangelicals see as central to their faith. This position also imagines that religious belief can be separated from other concerns. As we have seen, however, theology has never been separate from the desires and goals of the people who promulgate it. Hamilton is, however, right in another sense: white Evangelicals have transformed their theological commitments into political justifications that have affected and will continue to affect public policy on race and reproduction.

NOTES

1. Colin Kidd, *The Forging of Races: Race and Scripture in the Protestant Atlantic World, 1600–2000* (Cambridge: Cambridge University Press, 2006), 75.

2. Terence Keel, *Divine Variations: How Christian Thought Became Racial Science* (Stanford, CA: Stanford University Press, 2018), 6.

3. Kidd, *Forging of Races*, 2.

4. Camisha A. Russell, *The Assisted Reproduction of Race* (Bloomington: Indiana University Press, 2018), 2.

5. Michael O. Emerson and Christian Smith, *Divided by Faith: Evangelical Religion and the Problem of Race in America* (New York: Oxford University Press, 2001), 7.

6. Jemar Tisby, *The Color of Compromise: The Truth about the American Church's Complicity in Racism* (Grand Rapids, MI: Zondervan Reflective, 2019), 176.

7. Greg Smith, "More White Americans Adopted Than Shed Evangelical Label during Trump Presidency, Especially His Supporters," Pew Research Center, September 15, 2021, https://pewresearch .org/fact-tank/2021/09/15/more-white-americans-adopted-than-shed-evangelical-label-during -trump-presidency-especially-his-supporters/.

8. Joseph L. Graves Jr., "Out of Africa: Where Faith, Race, and Science Collide," in *Critical Approaches to Science and Religion*, ed. Myrna Perez Sheldon, Ahmed Ragab, and Terence Keel (New York: Columbia University Press, 2023), 255–76.

9. Nathaniel Jeanson, *Traced: Human DNA's Big Surprise* (Green Forest, AZ: New Leaf Publishing Group, 2022), 37.

10. Ken Ham and Charles Ware, *One Race One Blood (Revised & Updated): The Biblical Answer to Racism* (Green Forest, AZ: New Leaf Publishing Group, 2019).

11. Graves, "Out of Africa," 256.

12. Christopher P. Scheitle and Katie E. Corcoran, "COVID-19 Skepticism in Relation to Other Forms of Science Skepticism," *Socius* 7 (2021): 1–12.

13. C. Funk and B. Kennedy, "The Politics of Climate," Pew Research Center, October 4, 2016, https://pewresearch.org/internet/wp-content/uploads/sites/9/2016/10/PS_2016.10.04_Politics-of -Climate_FINAL.pdf.

14. Michael Lipka and Patricia Tevington, "Attitudes about Transgender Issues Vary Widely among Christians, Religious 'Nones' in U.S.," Pew Research Center, July 7, 2022, https://pewresearch.org/short -reads/2022/07/07/attitudes-about-transgender-issues-vary-widely-among-christians-religious -nones-in-u-s.

15. Kim Parker, Juliana Menasce Horowitz, and Anna Brown, "Americans' Complex Views on Gender Identity and Transgender Issues," Pew Research Center, June 28, 2022, https://pewresearch.org /social-trends/2022/06/28/americans-complex-views-on-gender-identity-and-transgender-issues.

16. Lawrence Ware, "If Your Pastor Says, 'Racism Isn't a Skin Problem, It's a Sin Problem,' You Need to Find Another Church," *The Root*, August 16, 2016, https://www.theroot.com/if-your-pastor-says-racism-isn-t-a-skin-problem-it-s-1822522858.

17. Ware, "If Your Pastor Says."

18. Michael O. Emerson, Christian Smith, and David Sikkink, "Equal in Christ, but Not in the World: White Conservative Protestants and Explanations of Black-White Inequality," *Social Problems* 46, no. 3 (1999): 398–417, 401.

19. Dorothy Roberts, *Shattered Bonds: The Color of Child Welfare* (London: Hachette UK, 2009).

20. Mark A. Noll, *God and Race in American Politics: A Short History* (Princeton, NJ: Princeton University Press, 2008), 29.

21. Noll, *God and Race in American Politics*, 29.

22. Kathryn Lofton, "Religious History as Religious Studies," *Religion* 42, no. 3 (2012): 383–94, 391.

23. Melissa J. Wilde, *Birth Control Battles: How Race and Class Divided American Religion* (Oakland: University of California Press, 2019), 4.

24. Rebecca Anne Goetz, *The Baptism of Early Virginia: How Christianity Created Race* (Baltimore: Johns Hopkins University Press, 2016), 11.

25. Angie Maxwell, Pearl Ford Dowe, and Todd Shields, "The Next Link in the Chain Reaction: Symbolic Racism and Obama's Religious Affiliation," *Social Science Quarterly* 94, no. 2 (2013): 321–43.

26. Goetz, *Baptism of Early Virginia*, 2.

27. Goetz, *Baptism of Early Virginia*, 63–64.

28. Goetz, *Baptism of Early Virginia*, 86.

29. Goetz, *Baptism of Early Virginia*, 86.

30. Goetz, *Baptism of Early Virginia*, 137.

31. Goetz, *Baptism of Early Virginia*, 137.

32. Goetz, *Baptism of Early Virginia*, 139.

33. Katharine Gerbner, *Christian Slavery: Conversion and Race in the Protestant Atlantic World* (Philadelphia: University of Pennsylvania Press, 2018); Heather Miyano Kopelson, *Faithful Bodies: Performing Religion and Race in the Puritan Atlantic* (New York: NYU Press, 2014).

34. Kathryn Gin Lum, *Heathen: Religion and Race in American History* (Cambridge, MA: Harvard University Press, 2022), 15–16.

35. Genesis 4:8–11, New American Standard Version.

36. Genesis 9:23.

37. Genesis 9:24–27.

38. Benjamin Braude, "The Sons of Noah and the Construction of Ethnic and Geographical Identities in the Medieval and Early Modern Periods," *William and Mary Quarterly* 54, no. 1 (1997): 103–42.

39. John Piper, "Red, Yellow, Black, and White—Could Every Race Come from Adam, Eve, and Noah?," *Desiring God*, August 26, 2016, https://desiringgod.org/interviews/red-yellow-black-and-white-could-every-race-come-from-adam-eve-and-noah.

40. Genesis 11:1–9.

41. Nyasha Junior, "The Mark of Cain and White Violence," *Journal of Biblical Literature* 139, no. 4 (2020): 661–73.

42. Terence D. Keel, "Religion, Polygenism and the Early Science of Human Origins," *History of the Human Sciences* 26, no. 2 (April 2013): 3–32, 5.

43. Kidd, *Forging of Races*, 62.

44. Kidd, *Forging of Races*, 64.

45. Goetz, *Baptism of Early Virginia*, 107.

46. Keel, "Religion, Polygenism and the Early Science," 8.

47. Keel, "Religion, Polygenism and the Early Science," 8.

48. Keel, "Religion, Polygenism and the Early Science," 9.

49. Richard H. Popkin, "Pre-Adamism in 19th Century American Thought: 'Speculative Biology' and Racism," *Philosophia* 8, no. 2–3 (1978): 205–39, 221.

50. Keel, *Divine Variations*, 78.

51. Kidd, *Forging of Races*, 149.

52. Kidd, *Forging of Races*, 148.

53. Christopher A. Luse, "Slavery's Champions Stood at Odds: Polygenesis and the Defense of Slavery," *Civil War History* 53, no. 4 (2007): 379–412.

54. Kidd, *Forging of Races*, 139.

55. To be clear, the slavery of the Hebrew Bible differed significantly from American chattel slavery. See Jonathan J. Hatter, "Slavery and the Enslaved in the Roman World, the Jewish World, and the Synoptic Gospels," *Currents in Biblical Research* 20, no. 1 (2021): 97–127; Tracy M. Lemos, "Physical Violence and the Boundaries of Personhood in the Hebrew Bible," *Hebrew Bible and Ancient Israel (HeBAI)* 2 (2013): 500–531; Stephen R. Haynes, *Noah's Curse: The Biblical Justification of American Slavery* (New York: Oxford University Press, 2002).

56. Luse, "Slavery's Champions Stood at Odds," 382.

57. Luse, "Slavery's Champions Stood at Odds," 380.

58. Luse, "Slavery's Champions Stood at Odds," 400.

59. Luse, "Slavery's Champions Stood at Odds," 409.

60. Jane Dailey, "Sex, Segregation, and the Sacred after *Brown*," *Journal of American History* 91, no. 1 (2004): 119–44, 123–24.

61. Genesis 6:11–22.

62. Kidd, *Forging of Races*, 150.

63. Kidd, *Forging of Races*, 138.

64. Ephesians 6:5; Colossians 3:22.

65. Kidd, *Forging of Races*, 140.

66. Kevin Jones, "Report on Slavery and Racism in the History of the Southern Baptist Theological Seminary," December 2018, https://digitalcommons.cedarville.edu/education_publications/108, 8.

67. Jones, "Report on Slavery and Racism," 55.

68. Jones, "Report on Slavery and Racism," 57.

69. Christine Rosen, *Preaching Eugenics: Religious Leaders and the American Eugenics Movement* (New York: Oxford University Press, 2004).

70. Rosen, *Preaching Eugenics*, 118.

71. Rosen, *Preaching Eugenics*, 120.

72. "American Eugenics Society #1," box 1, Charles B. Davenport Papers, American Philosophical Society, Philadelphia.

73. "American Eugenics Society #1."

74. Rosen, *Preaching Eugenics*, 141–42; Charles McDaniel, "John A. Ryan and the American Eugenics Society: A Model for Christian Engagement in the Age of Consumer Eugenics," *Journal of Religion and Society* 22 (2020): 1–22, 6.

75. Rosen, *Preaching Eugenics*, 158–61.

76. Rosen, *Preaching Eugenics*, 16–17.

77. Rosen, *Preaching Eugenics*, 17.

78. Rosen, *Preaching Eugenics*, 66.

79. Melissa J. Wilde, "Complex Religion: Interrogating Assumptions of Independence in the Study of Religion," *Sociology of Religion* 79, no. 3 (2018): 287–98, 289.

80. Edward J. Larson, *Sex, Race, and Science: Eugenics in the Deep South* (Baltimore: Johns Hopkins University Press, 1996), 9.

81. Sean McCloud, *Divine Hierarchies: Class in American Religion and Religious Studies* (Chapel Hill: University of North Carolina Press, 2009), 35.

82. *Brown v. Board*, 347 U.S. 483 (1954).

83. J. Russell Hawkins, *The Bible Told Them So: How Southern Evangelicals Fought to Preserve White Supremacy* (New York: Oxford University Press, 2021).

84. Hawkins, *Bible Told Them So*, 21.

85. Hawkins, *Bible Told Them So*, 44.

86. Karen S. Anderson, "Massive Resistance, Violence, and Southern Social Relations: The Little Rock, Arkansas, School Integration Crisis, 1954–1960," in *Massive Resistance: Southern Opposition to the Second Reconstruction*, ed. Clive Webb (New York: Oxford University Press, 2005), 203–19, 211.

87. Charles J. Ogletree Jr., "The Significance of *Brown*," *Judicature* 88 (2004): 66–72.

88. Hawkins, *Bible Told Them So*, 148.

89. *Newman v. Piggie Park Enterprises, Inc.*, 390 U.S. 400 (1968).

90. Steve Suitts, *Overturning Brown: The Segregationist Legacy of the Modern School Choice Movement* (Montgomery: New South Books, 2020), 12.

91. Hawkins, *Bible Told Them So*, 149.

92. *Runyan v. McCary*, 427 U.S. 160 (1976).

93. *Bob Jones University v. United States*, 461 U.S. 574 (1983).

94. Hawkins, *Bible Told Them So*, 162.

95. Kristin Kobes Du Mez, *Jesus and John Wayne: How White Evangelicals Corrupted a Faith and Fractured a Nation* (New York: Liveright, 2020), 157.

96. Emerson and Smith, *Divided by Faith*, 88.

97. Kobes Du Mez, *Jesus and John Wayne*, 157.

98. Tisby, *Color of Compromise*, 152.

99. Kobes Du Mez, *Jesus and John Wayne*, 71.

100. The general scientific consensus is that humans arrived in Australia around 50,000 years ago and people arrived in the Americas somewhere between 15,000 and 30,000 years ago.

101. John MacArthur, "The Sin of Noah," Grace to You, June 17, 2001, accessed August 19, 2023, https://www.gty.org/library/sermons-library/90-264/the-sin-of-noah.

102. Ryan P. Burge and Miles D. Williams, "Is Social Media a Digital Pulpit? How Evangelical Leaders Use Twitter to Encourage the Faithful and Publicize Their Work," *Journal of Religion, Media and Digital Culture* 8, no. 3 (2019): 309–39, 316.

103. Piper, "Red, Yellow, Black, and White."

104. Piper, "Red, Yellow, Black, and White."

105. Genesis 6:5–12.

106. Albert Mohler, "The Briefing," August 30, 2019, https://albertmohler.com/2019/08/30/briefing-8-30-19.

107. Mohler, "Briefing," August 30, 2019.

108. Mohler, "Briefing," August 30, 2019.

109. "The Answers," Billy Graham Evangelistic Association, March 6, 2020, https://billygraham.org/answer/can-behavior-be-changed-by-altering-genes/.

110. Albert Mohler, "Is Homosexuality in the Genes?," April 4, 2004, https://albertmohler.com/2004/04/15/is-homosexuality-in-the-genes.

111. Mohler, "Briefing," August 30, 2019.

112. Jeff Johnston, "Male and Female He Created Them: Genesis and God's Design of Two Sexes," *Focus on the Family*, September 13, 2015; Christopher Yuan, "He Made Them Male and Female: Sex, Gender, and the Image of God," *Desiring God*, December 14, 2019, https://desiringgod.org/articles/he-made-them-male-and-female.

113. Steven Wedgeworth, "The Science of Male and Female: What God Teaches Through Nature," *Desiring God*, September 11, 2020, https://desiringgod.org/articles/the-science-of-male-and-female.

114. "Nashville Statement," Coalition for Biblical Sexuality, accessed July 24, 2023, https://cbmw.org/nashville-statement/.

115. Albert Mohler, "The Briefing," June 17, 2022, https://albertmohler.com/2022/06/17/briefing-6-17-22.

116. Emerson et al., "Equal in Christ," 401.

117. Emerson and Smith, *Divided by Faith*, 116.

118. Emerson et al., "Equal in Christ," 401.

119. For conservative white Evangelical Christians, within this complementary relationship there is a hierarchy structured by "the loving, humble leadership of redeemed husbands and the intelligent, willing support of that leadership by redeemed wives." See Wayne Grudem and John Piper, "The Danvers Statement: Recovering Biblical Manhood and Womanhood: A Response to Evangelical Feminism," Council on Biblical Manhood and Womanhood, 1997.

120. Du Mez, *Jesus and John Wayne*, 38–39.

121. Eduardo Bonilla-Silva, *Racism without Racists: Color-Blind Racism and the Persistence of Racial Inequality in the United States* (Lanham, MD: Rowman and Littlefield, 2006), 2.

122. Evangelical Protestants have explicitly linked marriage and patriotism. For example, the Family Leaders' "Marriage Vow" asks candidates to vow "personal fidelity to my spouse" and "official fidelity" to the US Constitution and claims that some judges who "have personally rejected heterosexuality and faithful monogamy, have also abandoned bona fide constitutional interpretation in accord with the discernible intent of the framers." "The Marriage Vow," Wikipedia, accessed April 12, 2024.

123. James N. Gregory, *The Southern Diaspora: How the Great Migrations of Black and White Southerners Transformed America* (Chapel Hill: University of North Carolina Press, 2005), 316.

124. Omri Elisha, "Moral Ambitions of Grace: The Paradox of Compassion and Accountability in Evangelical Faith-Based Activism," *Cultural Anthropology* 23, no. 1 (2008): 154–89; Brantley W. Gasaway, *Progressive Evangelicals and the Pursuit of Social Justice* (Chapel Hill: University of North Carolina Press, 2014).

125. Albert Mohler, "The Heresy of Racial Superiority—Confronting the Past, and Confronting the Truth," June 23, 2015, https://albertmohler.com/2015/06/23/the-heresy-of-racial-superiority-confronting-the-past-and-confronting-the-truth.

126. Rick Warren, "How to Overcome the Sin of Prejudice," June 15, 2020, https://pastorrick.com/how-to-overcome-the-the-sin-of-prejudice/.

127. Franklin Graham, "The Love That Conquers Racism," *Decision Magazine*, July 1, 2020, https://decisionmagazine.com/franklin-graham-the-love-that-conquers-racism/.

128. John MacArthur, "Act like Men," Grace to You, June 21, 2020, https://www.gty.org/library/sermons-library/81–82/act-like-men.

129. Samuel Sey, "Murders That Won't Go Viral: The Quiet Injustice Too Few Protest," *Desiring God*, October 30, 2020, https://desiringgod.org/articles/murders-that-wont-go-viral.

130. MacArthur, "Act like Men."

131. Michalinos Zembylas, "Affect, Race, and White Discomfort in Schooling: Decolonial Strategies for 'Pedagogies of Discomfort,'" *Ethics and Education* 13, no. 1 (2018): 86–104; Megan R. Underhill, "'Diversity Is Important to Me': White Parents and Exposure-to-Diversity Parenting Practices," *Sociology of Race and Ethnicity* 5, no. 4 (2019): 486–99.

132. Albert Mohler, "The Briefing," June 24, 2020, https://albertmohler.com/2020/06/24/briefing-6-24-20.

133. Mark Driscoll, "What Does Depravity Mean and Not Mean?" Real Faith by Mark Driscoll, accessed August 15, 2023, https://realfaith.com/daily-devotions/what-does-total-depravity-mean-and-not-mean/.

134. Timothy Keller, "The Bible and Race," *Life in the Gospel*, Spring 2020, https://quarterly.gospelinlife.com/the-bible-and-race/.

135. John Piper, "Structural Racism: The Child of Structural Pride," *Desiring God*, November 15, 2016, https://desiringgod.org/articles/structural-racism.

136. Mohler, "Briefing," June 24, 2020.

137. Mohler, "Briefing," June 24, 2020.

138. MacArthur, "Act like Men."

139. Mohler, "Briefing," June 24, 2020.

140. Mohler, "Briefing," June 24, 2020.

141. Mohler, "Briefing," June 24, 2020.

142. Piper, "Structural Racism."

143. Timothy Keller, "The Sin of Racism," *Life in the Gospel*, Summer 2020, https://quarterly .gospelinlife.com/the-sin-of-racism/.

144. Graham, "Love That Conquers Racism."

145. Keller, "Bible and Race."

146. Driscoll, "What Does Depravity Mean?"

147. Driscoll, "Church: What Are the Characteristics of the Church?"

148. Rick Warren, "You're Called to Belong to the Church," August 22, 2019, https://pastorrick .com/youre-called-to-belong-to-the-church/.

149. Emerson et al., "Equal in Christ," 401.

150. Emerson et al., "Equal in Christ," 401.

151. Emerson et al., "Equal in Christ," 414.

152. Jesse Curtis, *The Myth of Colorblind Christians: Evangelicals and White Supremacy in the Civil Rights Era* (New York: NYU Press, 2021), 171.

153. Sharan Kaur Mehta, Rachel C. Schneider, and Elaine Howard Ecklund, "'God Sees No Color' So Why Should I? How White Christians Produce Divinized Colorblindness," *Sociological Inquiry* 92, no. 2 (2022): 623–46, 635.

154. "Religion and Congregations in a Time of Social and Political Upheaval," Public Religion Research Institute, May 16, 2023, https://prri.org/research/religion-and-congregations-in-a-time-of -social-and-political-upheaval/.

155. Kevin D. Dougherty, Mark Chaves, and Michael O. Emerson, "Racial Diversity in U.S. Congre- gations, 1998–2019," *Journal for the Scientific Study of Religion* 59, no. 4 (2020): 651–62, 660.

156. "Stepping into a Multi-Cultural World: Racial Reconciliation and the Gospel," Lifeline Chil- dren's Services, accessed August 18, 2023, https://lifelinechild.org/stepping-into-a-multi-cultural -world-racial-reconciliation-the-gospel/.

157. "Stepping into a Multi-Cultural World."

158. Samuel L. Perry, *Growing God's Family: The Global Orphan Care Movement and the Limits of Evangelical Activism* (New York: NYU Press, 2017).

159. Samuel L. Perry and Andrew L. Whitehead, "Christian Nationalism, Racial Separatism, and Family Formation: Attitudes toward Transracial Adoption as a Test Case," *Race and Social Problems* 7, no. 2 (2015): 123–34, 134.

160. Perry and Whitehead, "Christian Nationalism," 124.

161. Risa Cromer, "'Our Family Picture Is a Little Hint of Heaven': Race, Religion and Selective Reproduction in U.S. 'Embryo Adoption,'" *Reproductive Biomedicine and Society Online* 11 (2020): 9–17.

162. Tisby, *Color of Compromise*, 175.

163. Tisby, *Color of Compromise*, 153.

164. Randall Balmer, *Bad Faith: Race and the Rise of the Religious Right* (Grand Rapids, MI: Eerd- mans, 2021).

165. Balmer, *Bad Faith*, 68.

166. Balmer, *Bad Faith*, 67.

167. Robert P. Jones, *White Too Long: The Legacy of White Supremacy in American Christianity* (New York: Simon and Schuster, 2021), 6.

168. Anthea Butler, *White Evangelical Racism: The Politics of Morality in America* (Chapel Hill: University of North Carolina Press, 2021).

169. Daniel Cox and Robert P. Jones, "Deep Divide between Black and White Americans in Views of Criminal Justice System," Public Religion Research Institute, May 7, 2015, https://prri.org/research /divide-white-black-americans-criminal-justice-system/.

170. Diana Orcés, "Black, White, and Born Again: How Race Affects Opinions among Evangelicals," Public Religion Research Institute, February 17, 2021, https://prri.org/spotlight/black-white-and -born-again-how-race-affects-opinions among evangelicals/.

171. "Summer Unrest over Racial Injustice Moves the Country, but Not Republicans or White Evangelicals," Public Religion Research Institute, August, 21, 2020, https://prri.org/press-release/summer -unrest-over-racial-injustice-moves-the-country-but-not-republicans-or-white-evangelicals/.

172. Emerson and Smith, *Divided by Faith*, 170.

173. Michael S. Hamilton, "A Strange Love? Or: How White Evangelicals Learned to Stop Worrying and Love the Donald," in *Evangelicalism, Its History, and Its Current Crisis*, ed. David Bebbington, George Marsden, and Mark Noll (Grand Rapids, MI: Eerdmans, 2018), 221.

7

How Does a Baby Have a Race?

Alice B. Popejoy

ROYAL RACISM

The Netflix series *Harry & Meghan* is an insider exposé of US-born Meghan Markle and her now husband, Prince Harry, the Duke of Sussex, sharing their experiences navigating internal politics and the British tabloids' portrayal of Meghan as a caricature of negative racial stereotypes.[1]

When Meghan and Harry first publicly revealed their relationship, she became a prospective member of the British royal family, and public favor was generally with the young couple, bolstered by Meghan's popularity as an amiable American actress who starred on the TV series *Suits*.[2] Once they became engaged and Meghan's popularity grew across the Commonwealth while on tour with Harry, subjects in parts of the Global South still under the British Crown suddenly felt they could relate to a royal. While those at Buckingham Palace may have quietly questioned the acceptability of an actress marrying a prince, the reception of their engagement in the United Kingdom was generally positive.

A beautiful American actress who seemed to effortlessly win the affection of the Queen of England (Harry's grandmother), Meghan was praised as a highly suitable girlfriend for the handsome prince. The media commented on her sharp sense of style, graceful ease, and natural warmth in front of the camera. Suggesting she might be "the one," pop culture magazines speculated about the couple's seriousness, considering Harry's string of prior romantic relationships and highly publicized escapades.[3]

The Netflix series reveals that the couple started dating seriously out of public view, keeping their relationship mostly hidden in private residences and telling only a trusted few. They became engaged not long after going public, which felt sudden to some. As the prince's fiancée, Meghan came under more scrutiny than she had as his girlfriend. Suddenly, she was no longer judged for her appearance, affect, and profession; rather, speculations emerged about her role within the royal family. This change in tone invoked racism, as some suggested that Meghan, the daughter of a white father and a Black mother, would not "fit in" at the palace.

Meghan was taken aback when the UK-based gossip columns and newspaper articles began blasting her (now Prince Harry's fiancée) in a negative way while invoking racial stereotypes of Black Americans. As this book's chapters by Mark Fedyk and Lisa Ikemoto explain, identity is complex. On camera for Netflix, Meghan described not having previously identified as a "Black woman" in the United States because she had light skin and was racially ambiguous. Meghan and her father, a white man, were close throughout her childhood. She characterized herself as a "Daddy's girl" and seemed to relate more to his racial identity than to her mother's. Their previous closeness was a source of pain and distress when Meghan and her father became estranged after he sold fake stories to the relentless British tabloids. In the series, Meghan explains how she had not much pondered having differently racialized parents until it became a focus on social media and in the UK popular press. Harry, too, isolated and protected from awareness of modern-day racism by his white (and royal) privilege, received a swift lesson in its harmful effects.

Meghan's mother, a self-described Black woman with darker skin than her daughter, was often mistaken as a nanny or the adoptive mother of her biological child due to the differences in their appearance. In contrast to Meghan's and Harry's surprise at the press invoking racist stereotypes, Meghan's mom saw it coming. Having lived in the United States as a Black woman her entire life, she spoke on camera about having latent anxiety that racism would eventually harm her daughter.

Leading up to the wedding, Meghan was judged with increasing British scrutiny. Harassment from the paparazzi grew to an unbearable extreme, not unlike the days of Princess Diana (the late mother of Harry and William), when a royal jewel on her ring finger marked the end of a personal life and the freedom to live as an independent woman in the world.[4] As the UK public(s) and the royal family began to see these women as future assets of the Crown—no longer simply young, beautiful women winning the hearts of princes—their unique identities and bright characters seemed to fade, being gradually replaced by distorted daily narratives and motives of the popular press.

In Meghan's case, racialization of her identity made the prospect of becoming Prince Harry's wife a controversial one. Sixty years ago, marriage between people assigned to different race categories was illegal in many US states, and it may still be frowned upon in some families and communities. With an undertone of race-based disapproval, the British media portrayed Meghan as a questionable reproductive contributor to the royal bloodline, given her familial heritage. Even before she and Harry officially tied the knot, Meghan's prospective membership in the royal family became a topic of racialized debate on reproduction.[5]

Racist media attacks escalated in frequency and intensity when Meghan became pregnant after the wedding and was compared negatively to Harry's older brother William's wife, Kate (formerly Middleton), the Princess of Wales. Photos of the two women holding their pregnant bellies in the same ways (at different times) touted Kate's protective motherly instincts, while accusing Meghan of being obsessed with her baby bump, reportedly keeping her hand there as a means of gaining attention. While neither woman came from a royal background, Kate's (entirely) white British background made her an appropriate vessel for the reproduction of royalty, while Meghan's mixed racial background made her a dubious one.

In a cruel turn, British tabloids narrowed in on the predicted race of the embryo in Meghan's womb. The Netflix exposé implies that members of the royal family (presumably William and the brothers' father, Charles) had expressed concern for the future of the British monarchy should the child be born "too dark" for their standards. Excruciating paparazzi harassment, controlling policies of the royal family's administration, and threats of physical harm led Meghan into devastating isolation and depression during her pregnancy. Harry declined to comment on the specifics but made it clear that he and Meghan were forcibly ousted from the royal family, while media narratives maintained that they left by choice. The stress from these circumstances undoubtedly influenced Meghan's mental and physical health, and thus (potentially) the health of her future baby boy, Archie.

The publicity surrounding Meghan and Harry's marriage and the birth of their first child throws into sharp relief the usually implicit existential threat that reproduction among racially discordant partners poses to white supremacy and, as Lisa Ikemoto's chapter points out, the notion of white purity, which are both foundational principles of colonialism. The harmful ideologies of white supremacy and purity are critical mechanisms of the British monarchy's persistence, as they continue to justify global exploitation and oppression. Colonialism's justification has always been rooted in the premise that a small number of (divinely ordained) people deserve rights to resources and rulership that extend across the globe, while others are meant to toil in poverty without access to even their own lands' riches. Thus, reproduction between favored and unfavored groups naturally invalidates the falsehoods perpetually touted about biological underpinnings of human classes. The same is true of mixed-race children; their very existence threatens the idea that humans can be distinctly divided into coherent racial groups. These offspring are not easily placed into any of the mutually exclusive race-labeled categories, thus gradually dissolving any imaginary dividing lines between them.

Though *Harry & Meghan* made headlines for its raw and intimate exploration of the couple's personal experiences in a blinding, frightening spotlight, it also shed light on an important phenomenon for science, medicine, and public policy that has yet to be deeply interrogated: the construction and attribution of race categories to entities with no ability or agency to participate in or influence this process, such as a developing embryo, a newborn baby, or a young child.

RACIALIZABILITY

Considering the race of a baby would not be so challenging if race were an inherent biological trait that could be determined with a scientific approach. If race could be measured objectively, with classification rules uniformly applied, individuals could be unambiguously assigned to groups. It is precisely because race is socially constructed, and because categories are ephemeral, subjective, and context dependent, that it is extremely difficult to answer the question of whether a baby can have a race at all.[6]

If an entity can be presumed to belong to a racial category despite its inability to persist outside of the womb, how far does the attributability of race extend back in the growth of a budding fetus? Would a clump of cells in the earliest stages of embryonic development qualify as racializable? If so, at what stage of embryogenesis would it be feasible (and socially acceptable) to speculate about the embryo's racial identity, and, if based on the racial identities of its parents, would every potential offspring of a certain reproductive pairing be predicted to have the same race?

Every adult human being has developed personal beliefs about racial identity and reproduction, whether consciously or not, at the very least related to their own birth parents and their social or cultural heritage. It is almost certainly the case that, for some reproductive scenarios, it may be difficult to imagine how a baby could have a different racial identity than that of its parents—for example, when the two parents have similar racial identities. When parents' racial identities differ, this may be a more interesting question with no concrete answers. In either scenario, it may be tempting to assert with confidence the predicted race of a baby still in utero, but we must consider the implications of this assertion.

Biomedical researchers are engaged in the creation of human cell lines—that is, cell cultures derived from human cells, which are extracted and then grown in a laboratory. Do these cell lines maintain the racial identities of the humans from whom they were derived? If so, is there a rigorous way to take this into account when conducting genetic or other health-related scientific research using human cell lines? And how might we interpret our findings, if indeed such qualities were identified and associated with interesting biomarkers?

If a cell can have a racial identity, this could fundamentally change the definition of race from a thing social scientists, anthropologists, and demographers tell us emerges through a socially and culturally derived process of self-identification and ascription that is fluid and dynamic across contexts to something that is fixed and biological in nature. Race might instead be treated as a heritable trait, transmitted genetically from parent to offspring, which persists in its potential relevance for biology even after it is removed from a human body and maintained in an artificial environment. But what if the "racial identity" of a cell lies not in its DNA but in its epigenome, the annotations that are added to our DNA in the form of methylation, acetylation, or changes to histone structure throughout our lives? The "racial identity" of the cell could then reflect the racialized experiences of the cell's donor, rather than any innate genetic factors.

Treating race as an immutable trait that nonhuman (cell lines) or future human (embryos) entities can possess is a slippery slope, whether a single cell in a petri dish or a growing fetus in the womb. In the case of Archie and his ex-royal mother, tabloid-fabricated racialized tropes about Meghan may have trickled down to influence Archie's development in negative ways, through the stress of growing a human being in a system that is designed to oppress, extract, and exploit. I therefore argue that it is not fabricated or assigned racial identities that alone do the work of racism. It is rather through the context-specific applications and uses of those labels that facilitate and enact race-based discrimination and aggression that harm is done to individuals and communities. In this line of thinking, race labels may or may not be assigned to babies, but this has no impact on them unless there is a conceptual framework and hierarchical systems of power that create meaning and produce harm (or protection) according to and abiding by those labels.

The act of racializing a person (or a thing) is assigning a label for the purpose of establishing or maintaining hierarchies of difference, and is an opportunity to arbitrarily generate or sequester power.[7] Interrogating how newborns are racialized, for what purpose(s), and by whom offers insights about the creation and reproduction of race and ethnicity labels as means of administrative, medical, and scientific classification, thereby undermining their scientific validity and clinical utility.

The central thesis of this chapter is that a baby does not have a fully formed cognitive or conscious racial identity, and that both direct and indirect assertions or assignments of attributed race are derivative and deterministic in nature. That is, they are derived from and play a part in determining racial constructs in a cyclical loop of self-referential and reinforcing logic. Racial labels are used to separate human beings into mutually exclusive categories of difference based on a handful of arbitrary physical characteristics and how they interact with the social, cultural, and political context at a particular time and place.[8] Systems in which people are inherently stratified according to these social structures will inevitably yield group-level differences that correlate with racial labels and thereby appear to naturalize racial categories.

This closed-loop system operates on both conscious and unconscious levels, with underlying systemic, economic, and political motivations that have been buried in historical narratives with which most scientists and physicians are unfamiliar or believe are irrelevant.[9] This system functions to preserve extant hierarchical structures in society, with the ephemeral attribute of whiteness (which many biomedical scientists and physicians in the United States and the United Kingdom share) as the ultimate gatekeeper and passport to power. Since the fundamental nature of this construct is self-reinforcing without revealing its foundational conceptual flaws, everyone must diligently engage in questioning its use in science and society.

ON WHITENESS

Humans interpret our experiences and observations through a sensory-neuronal process of creating meaning about inputs our brains receive through sense organs. We *Homo sapiens* are particularly reliant on our visual systems to provide information about our surroundings. We must render down a vast stream of input to a simplified and comprehensible pattern of thoughts and images to make sense of the world. This process has played an important role in helping humans to determine safety and kinship ties throughout our evolutionary history, and it has also contributed to our survival-based cultural notions of "other."[10]

Our natural way of processing information about the world and putting it into boxes—together with unhelpful human traits like greed and compartmentalization—influences our beliefs about what our observations might signify about a person, from biology to behaviors, and even the limits of one's potential across all domains of life. Specifically, the perception of an undefined trait—whiteness—suggests proximity to European dominance, power, and privilege.

One of the major ideological dogmas of the US political and social order is a belief in the concept of whiteness as indicative of innate deservedness, or a natural inheritance of unearned dominance, power, and privilege. In the natural and medical sciences, whiteness is rarely discussed as a scientific or even conceptual object, taken for granted as the reference category in statistical analyses comparing data grouped by race or ethnicity labels. It is implicitly treated as the presumed default, or the standard, for US society.[11]

Reflecting histories of discrimination and oppression, biostatistical comparisons between white and "non-white" groups that identify disparities reinforce misconceptions of the scientific validity of whiteness. When researchers make comparisons between groups considered socially or scientifically relevant, implicit assumptions of validity and mutual exclusivity of categories are often unquestioned. When data are missing on race or ethnicity, they may be inferred from other information without asking individuals directly; or they may be excluded entirely.[12]

Missingness of race and ethnicity data is not random. Categorical frameworks and discrete data structures used to racialize people ensure that those who identify as white have at least one relevant box to check, corresponding to a label that denotes the quality of whiteness. In contrast, categories that are often described singularly as "non-white" render differently in our data collection frameworks and contextual understandings. It is not sufficient to identify someone based on the absence of a quality (i.e., non-whiteness), so once someone is labeled or determined as *non-white*, there is a subsequent layer of racial classification with categorical constructs that may (or may not) be relevant to a person's identity. In the research context, people who identify with more than one category are often arbitrarily assigned to one or another—or to a multiracial category—by the investigators. Often, the only salient option for multiracial people is a checkbox labeled "other." For someone like Meghan Markle, her concurrent status as descended from both white and Black parents reveals a rupture in the fabric of assumptions that underlie these racialized frameworks and the construct of whiteness.

Race and racism are inseparable. If the concept of racial identity is meant to situate people within a social order or hierarchy of power, Meghan Markle's racial identity confuses that social order and challenges hierarchies of power, which exposes her to further racism. So, what about her offspring? Babies lack any power; they are at the mercy of caregivers to keep them safe and alive. For that reason, they experience racism primarily through the benefits or harms of their parents' and/or caregivers' racial and economic privileges or lack thereof. If perceived whiteness is a requirement to access those benefits and privileges, and all racial categorization is done in reference or relative to whiteness, it follows that a baby can have a race only when an external authority assigns it a label to denote whiteness or a lack thereof; and these labels endow the baby either with value and privilege, or with inherent disadvantage.

RACIALIZATION AS AN INTERNAL
AND EXTERNAL PROCESS

The identity aspect and semantic object "infant"—from the Latin *infantem*, or *one who cannot speak*—supersedes any racial identity a baby may have that would grant or revoke their power. Even when race labels are assigned to an infant, if race is a construct to identify people, assign stereotypes, and designate social position with its attending privileges (or lack thereof), then it is irrelevant to babies because the designation as "baby" takes precedence. This use of a semantic classifier denotes the total absence of power to determine one's own racial identity.

If we consider race as an internal process of integrating societal inputs into an aspect of one's personal identity, babies—and anyone else without a robust and developed sense of self—are ineligible as entities who could possibly have a racial or an ethnic identity. Viewed in this light, Meghan and Harry's unborn child, Archie, was not eligible for a racial identity, because his embryonic/fetal state rendered him incapable of complex internal processes of self-discovery. However, the racial identities of Harry and Meghan interacted with one another and with the sentiments of the British popular press to spark the imaginations of people who looked on with anticipation or anxiety about the future race of their baby.

Scholars across the social sciences and anthropology typically define race as a category of identity that is co-constructed between an individual and the society in which they live.[13] That is, our personal contexts and lived experiences shape

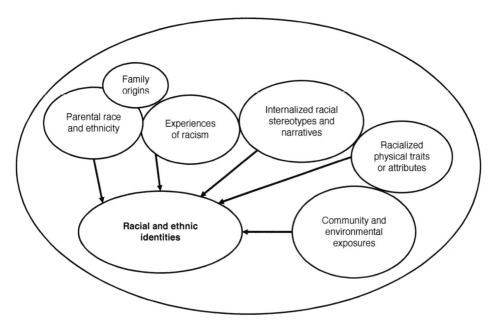

FIGURE 7.1. Multifaceted construction of race and ethnicity: context-dependent factors combine to create social, cultural, and political identities of race and ethnicity. Racial identities may be constructed through a complex process involving contextual understandings of oneself and one's family origins. Individual choices about how to self-identify are often constrained by predefined categories and stereotypes, and assignments of race categories are based on interwoven social, cultural, and political norms or conventions. Image created by the author.

our racial identities over time. The example of Meghan Markle illustrates how a person's racial identity can shift depending on the dynamics of their situation, which include public sentiments, popular press, societal narratives, and cultural norms. Figure 7.1 illustrates interactive processes through which self-identified race or ethnicity is developed in context.

The current "gold standard" for data collection on race and/or ethnicity is self-identification, also called self-reported race or ethnicity in the United States. Certificates of live birth for infants born in the United States have included information about race since 1916, when the standard certificate asked for the "color" of the mother and the father.[14] In 1997 the US Office of Management and Budget (OMB) revised its race-reporting standards, allowing people to indicate multiple races on the US Census.[15] Birth certificates have allowed parents to report multiple race categories only since 1999.[16] Revisions of and additions to US racial and ethnic categories and classification schemes have thus produced differences in the way demographic information has been recorded and characterized over time. The inherent fluidity of this data construct thus points to its fallacy as a proxy for something innate and immutable.

One might nevertheless encounter arguments and practices that suggest race or ethnicity exist objectively and could thus be assigned to individuals without their participation or agreement, including (presumably) babies. Babies may be profiled based on racial stereotypes of physical attributes (i.e., pigmentation of skin, hair, and eyes) and assigned a race label or category, for example. If race and ethnicity are thought of not as a process but rather as some natural essence that can be inherited through parents and can be objectively measured, ascertained by proxy or visual inspection and evaluation, then the absence of a personal, internal notion of one's own racial identity may not preclude babies from having race as an external characteristic or trait.

Scientists, clinical providers, and other medical or health-care professionals may implicitly (and incorrectly) consider race a biological essence that is passed on genetically from parent to child. Focusing on physical attributes of humans that have been socially and culturally coded as aspects of racial identity, people may reason that genetics are responsible for the physical development of such traits, and therefore that presumed racial attributes are inherited through DNA. While this interpretation of race is inconsistent with the contextually constructed view of race described in this volume, let us consider for a moment this position. Some physical traits are often attributed to stereotypical race categories, most notably relative concentrations of skin pigmentation but also hair texture and certain facial features.

The human pigmentation spectrum is determined by relative concentrations of melanin in our body's cells. Melanin is a protein produced in cells called melanocytes; it protects the body from radiation and helps with heat regulation through concentrations of pigmentation in skin, hair, and eyes.[17] Melanogenesis, the production of melanin, is controlled by innate genetic factors, as well as environmental factors such as cumulative exposure to UV light and changes in the body that are related to other gene-by-environment (GxE) processes such as inflammation. When babies are first born, the skin changes rapidly to create an epidermal barrier that protects the infant from harmful exposures in the world. This skin maturation process also involves significant changes in color, which increase with age.[18] Having two biological parents with different skin, hair, and eye colors introduces many different possibilities of coloration across physical traits that are traditionally racialized, which—by chance—may combine in ways that do or do not match stereotypical racial profiles in any given context.

Adopting a biological view of race based on an infant's relative concentrations of melanin would therefore be premature on two counts: (1) visual examination to assign race at birth preempts skin maturation with increasing baseline melanin in the first few weeks of life, and (2) environmental exposures trigger melanogenesis that darkens skin at different times, such as seasonally with increased daylight. The point here is to suggest not that racial or ethnic identities could ever be accurately surmised by external observers just by looking at an individual (because there is so much more to it—see figure 7.1) but rather that, even if one could determine someone's race by looking at them, newborn babies are not well suited to such methods of visual inspection and evaluation. This is, nonetheless, how race is often determined by morgue staff for the purpose of death certificates. In such instances, infants whose parents might not identify as white are often arbitrarily classified as white on postmortem inspection.

Babies born in the United States are not assigned a racial identity on their certificates of live birth based on visual inspection. In fact, they are not assigned a race at all. Birth certificates in the United States record only the racial and ethnic identities of the biological parents.[19] Since this information is self-reported, it meets the gold standard for race and ethnicity information, assuming that parents are given adequate opportunity to self-report. Subsequent users of birth certificates, however, including researchers as well as governmental and nongovernmental agencies, typically infer the race of the child from the race(s) of the parents. This practice exemplifies the assumption in US biomedicine, including research as well as clinical practice, that children must have the same racial identity as one or both parents.

RACE ASSIGNED BY PARENT PROXY

There are two basic approaches to assigning race using parent race and/or ethnicity as proxy variables: one method is to use data from just one parent, and the other is to use data from both parents. In the single-parent transmission model (figure 7.2-II-A), one categorical label is inherited from one parent, and nothing is inherited from the other parent. This model is used when the data for one parent is missing or unavailable, or when rules for coding race based on parental proxies indicate that a particular parent's labels should be transmitted preferentially. It may also be the result of discordant racial identities between parents, when one label is selected preferentially over another, regardless of the parent(s) from which they are inherited.

When race and ethnicity are concordant between parents and can be easily classified into one of the standard categories set forth by the OMB, the question of what race a prospective or actual newborn will be is considered straightforward. The baby will, it seems naturally, inherit its racial and ethnic identity from its parents. This practice breaks down, however, when parental race or ethnicity is ambiguous, when these identities differ between parents, or when one or both parents are missing data on race and ethnicity. Figures 7.2 and 7.3 illustrate models of transmission of parent race and ethnicity labels to infants when they are concordant, discordant, and/or missing.

Race label(s) assigned to an infant when parent labels are discordant or missing depend on predefined, structured rules for coding data into ontological frameworks, providing unique guidance for each context and instance in which race is assigned. Sometimes infants whose parents' race categories differ are assigned "other" or "multiracial" labels, whereas, in other settings, more complicated rules apply for selecting one category to represent an infant's race.

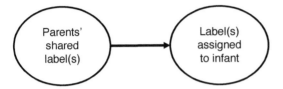

I. Concordant parent race and ethnicity labels
Parents have the same label(s) attributed to them, which are then assigned to the infant.

II. Discordant parent race or ethnicity labels
Parents have different labels attributed to them, so rules are applied to assign labels to the infant.

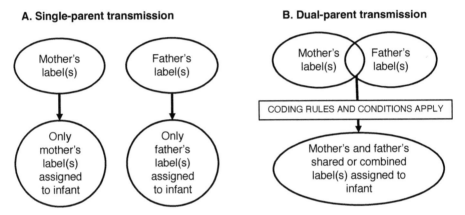

FIGURE 7.2. Models of transmission and assignment of categorical labels to infants based on parent race and ethnicity labels: (I) when parents' labels are concordant and (II) when parents' labels are discordant. (II-A) Single-parent transmission is implemented when either the mother or the father is designated as the parent whose racial and ethnic labels are to be assigned. (II-B) Dual-parent transmission is implemented when both parents' racial and ethnic categorical labels may be transmitted to their infant, based on different procedures or rules and conditions for coding this information. Image created by the author.

Figure 7.3 illustrates the hypothetical inference of infant race categories when parent race categories are discordant or missing. In figure 7.3-A, the mother is coded as "Hawaiian," and the father is coded as "Asian and Black." Under the rules of this hypothetical situation, since both parents are considered *non-white*, the father's race is transmitted to the infant, unless the mother is Hawaiian. In this case, then, the infant is also coded as "Hawaiian." When the mother is instead coded as "white," the default inheritance is the father's race because he is *non-white* (B and C). In the presence of multiple racial categories for a parent, the first *non-white* label listed is transmitted to the infant (i.e., "Asian" in B and "Black" in C). In

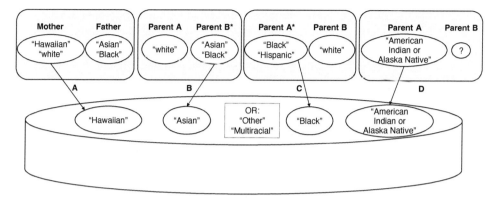

FIGURE 7.3. Hypothetical rules for how race labels could be assigned or attributed to infants based on parent labels. When parent labels are missing or discordant, various rules apply in different settings. Some labels are subjected to preferential transmission due to political or social dynamics; for example, only "Hawaiian" is assigned if the mother has this label (A). When one parent has been identified as "white" and the other parent has been identified with additional labels, only the first label attributed to the parent with non-"white" identities is transmitted (B, C). When only one parent has label(s), the first one listed is assigned to the infant (D). Image created by the author.

example D, only one parent's race label—"American Indian or Alaska Native"—is available, so this is transmitted without regard to the other parent.

This example of precoded algorithms for deriving infant race labels based on various combinations of parental labels shows how widely data on racial and ethnic self-identification may vary in provenance, structure, quality, and completeness. Missingness is nonrandom, and erroneous attribution is rampant. Even when categories do represent self-identified racial and ethnic identities, selecting more than one category often results in erasure of designations entirely, replacing them with a "mixed" or "biracial" or "two or more" category that represents *none* of the categories selected by a person (or assigned to them). Assigning race labels to babies by parent proxy is thus another example of racial designations being arbitrarily imposed.

RACE AS AN OBJECT OF SCIENTIFIC INQUIRY

When reporting to the National Institutes of Health (NIH) on the diversity of cell lines and samples of biomaterial collected and/or intended for use in federally funded research, investigators are compelled to report this information according to US Census categories. This requirement was congressionally mandated starting in 1993 to facilitate tracking and enhancing the diversity of participants in biomedical research. An unintended consequence of this requirement was reinforcing notions of mutually exclusive categories for race and ethnicity concepts among NIH-funded investigators, which likely motivated reliance on such categories for stratified analyses.[20]

There are many well-intentioned researchers who focus on health equity for racial and ethnic groups traditionally underrepresented in biomedical research, who are also underserved in medicine and by social support systems. Research that illuminates disparities in access to health care, outcomes tied to diminished services or reduced quality of care, and other types of analyses can help shed light on opportunities to make medicine more equitable. In these cases, it may be useful to rely on race and ethnicity categories.

There may, however, be negative consequences of publishing research and media articles that report on differences between people attributed to different race and ethnicity categories, because these narratives reinforce false notions of biological or genetic distinctions, such as those described in the chapter by Tina Rulli and in the conclusion to this volume. One example is reporting on US Black maternal and infant mortality associated with birth. When statistics such as infant mortality rates are presented as differences between Black and white infants (rather than the differences between babies of Black and white mothers), readers may implicitly assume (or, worse, the discussion sections may suggest) that there is something fundamentally wrong with Black infants—that they are more likely to die because of an underlying condition—rather than see it as an indicator of the physiological stress created by the mother's lived experience of racism, disparities in their mothers' access to prenatal care, or a reduced level of quality in prenatal and birthing care relative to white families.

A systematic review of articles published in US perinatal health-care research between 1980 and 2021 examined concordance among different race and ethnicity data collection techniques for infants.[21] The review authors concluded that "infants of color and those born to racially and/or ethnically discordant parents were the most likely to be misclassified across data sources" and that this misclassification of infants leads to inaccuracies in the measurement and reporting of morbidity and mortality rates across racial and ethnic categories. Study results indicate underestimates of these measures in "minoritized populations" relative to the "non-Hispanic/Latinx white population."

While this systematic review provides useful insights into the ways in which data collection and racial classification procedures create data disparities, with downstream implications for research findings and population-level estimates of important health metrics, it also rests on implicit (yet questionable) assumptions about the baseline validity of racial and ethnic classification. Authors acknowledged the "complicated" nature of collecting self-reported race and ethnicity data about infants due to their inability to self-report, and the lack of standards for collecting these data. They nevertheless evaluated different measurements of race and ethnicity for concordance, interpreting differences as *inaccurate*.

To say that a metric of inquiry is measured *inaccurately* means that it has a presumed ground truth to begin with; in this case, investigators treated parents' birth certificate race and ethnicity as the ground truth, and discordant race and ethnicity data on death certificates as misclassifications. Even in this study that was carefully designed to unpack limitations of racial and ethnic categorization techniques, race realism still underlies the entire analytic approach.

Decisions made in the realms of biomedical research, study design, data cleaning and analysis, and publishing results all add up to influence public (as well as scientific and medical) perceptions about the role of race and ethnicity in health and disease. This can sometimes have serious consequences for patients whose doctors make decisions or judgment calls rooted in common misconceptions about race. The onus is on human subject researchers to question and disrupt practices that are fundamentally unsupported as having scientific or clinical validity.

Questions about whether (or not) a baby can have a race should focus on the ways in which race realism is perpetuated, and on the data structures and analyses that maintain current practices to support the status quo. Rather than ask "How does a baby have a race?" perhaps the more precise question to ask is "Why do we think a baby should have a racial identity, and what is the validity of instruments we would use to attribute race to a baby?"

CONCLUSION: POWER AND POSITIONALITY

Racial ideologies underpin all individual and systemic applications of racism, from the personal development of racial identities to collective campaigns for data generation on different groups of people as defined by racial categories and frameworks. While US state and federal laws have eradicated many (explicitly) racist policies, and others have been rendered invalid due to greater protections for human rights and civil liberties, US infrastructures and the social, cultural, and political landscape of society have inherently disadvantaged individuals and communities that are classified or perceived as anything other than white.

This racial binary provided a justificatory foundation for Western imperialism, colonization, and their attending harms across the globe. Careful attention to and insistence upon empirical evaluation of the validity of such concepts is therefore required. The British monarchy enacted a presumed divine right to rule over previously sovereign nations; to oppress Indigenous, Native, and First Nations peoples; to extract and appropriate natural resources; and to justify colonial-exploitive historical practices under current global hegemonies based on an inherited endowment, supposedly bestowed by God. Religious depictions of light-skinned deities, contrasted with sinful humans shrouded in darkness, reinforce unconscious beliefs about lightness denoting *good* and darkness denoting *evil*, a trope that played out in the British tabloids and ultimately resulted in Meghan and Harry's separation from the royal family.

If even a *potentially Black* British prince-to-be poses a threat to the Crown's authority over populations who had never previously seen themselves represented among members of the royal family, does that not suggest the royal family's reliance on implicit whiteness to maintain its imperial supremacy?

This chapter has called into question the underlying assumptions, conceptual and scientific validity, and utility of measures intended to represent the race or ethnicity of an unborn fetus or a newborn baby. The ways in which US racial identities are constructed and attributed to those who lack agency to self-identify have been illuminated, and whiteness has been explicitly examined as a means of exclusion and erasure of complex identities. When the racial identity that one has grown up developing (often unconsciously) as a core sense of how one relates to the world has never been a source of pain, suffering, or concern, it is likely to remain unnoticed. This was illustrated with emotional sincerity and profound self-reflection in *Harry & Meghan* when Prince Harry revealed the depth of his privileged ignorance around racial politics, and how deeply ingrained they are in service to the royal British Empire turned Commonwealth. As Harry and Meghan's son, Archie, grows up and becomes self-aware within this global and historical context, he will have to reconcile his own conflicting identities the way his mother Meghan has done and continues to do, and as all children must do when their parents have different racial identities.

Without greater racial and ethnic diversity among scientific researchers in an academic field, blind spots and assumptions shared by people whose lives are not negatively impacted by attributed race will continue to creep in, causing errors in study design and interpretation that may lead to immediate and downstream harms.[22] Becoming aware of this is (for some) the beginning of a journey toward self-discovery and self-determination within historical, cultural, and societal contexts. For white scientists and physicians, this may be the beginning of a journey toward understanding how privilege and positionality have shaped them; and it is a call to wield the power endowed by whiteness in the United States and the United Kingdom to raise awareness and change the status quo.

Babies may be labeled or assigned racial identities when they lack power to influence this attribution, but as they develop into self-reflective individuals with agency and a will to express their own process of racial identity formation, they can claim that power to self-identify.

NOTES

1. Liz Garbus, dir., *Harry & Meghan* (Netflix, 2022), https://netflix.com/title/81439256.

2. Aaron Korsh, creator, *Suits* (Netflix, 2011–19), https://imdb.com/title/tt1632701.

3. Marina Hyde, "Meghan Markle Is the Perfect Fit for Our New Touchy-Feely Royal Family," *Guardian*, September 7, 2017, https://www.theguardian.com/lifeandstyle/lostinshowbiz/2017/sep/07/meghan-markle-perfect-fit-touchy-feely-royal-family-prince-harry.

4. Michael Holden, "Harry and Meghan Decry 'Pain and Suffering' of Women Brought into UK Royal Family," Reuters, December 8, 2022, https://www.reuters.com/lifestyle/british-royals-brace-harry-meghans-netflix-broadside-2022-12-08/.

5. Georgina Lawton, "I Usually Disparage the Royals, but Meghan Markle Has Changed That," *Guardian*, October 21, 2017, https://www.theguardian.com/lifeandstyle/2017/oct/21/royals-meghan-markle-uk-change-race-relations-prince-harry/; Nadifa Mohamed, "As Meghan Has Learned, the Monarchy Is Still Built on Breeding, Ancestry and Caste," *Guardian*, March 8, 2021, https://theguardian.com/commentisfree/2021/mar/08/meghan-monarchy-breeding-ancestry-caste.

6. C.C. Gravlee, "How Race Becomes Biology: Embodiment of Social Inequality," *American Journal of Physical Anthropology* 139, no. 1 (2009): 47–57.

7. Gail C. Christopher, Nikole Hannah-Jones, and Derrick Johnson, "Racial Hierarchy, Race Narrative, and the Structures That Sustain Them," in *Necessary Conversations: Understanding Racism as a Barrier to Achieving Health Equity*, ed. Alonzo L. Plough (New York: Oxford University Press, 2022), 19–33.

8. Audrey Smedley, *Race in North America: Origin and Evolution of a Worldview* (New York: Routledge, 2018).

9. Vence L. Bonham, Sherrill L. Sellers, Thomas H. Gallagher, Danielle Frank, Adebola O. Odunlami, Eboni G. Price, and Lisa A. Cooper, "Physicians' Attitudes toward Race, Genetics, and Clinical Medicine," *Genetics in Medicine* 11, no. 4 (2009): 279–86.

10. Karen L. Kramer, "The Human Family—Its Evolutionary Context and Diversity," *Social Sciences* 10, no. 6 (2021), https://doi.org/10.3390/socsci10060191.

11. Michael Omi and Howard Winant, *Racial Formation in the United States* (New York: Routledge, 2014).

12. Alice B. Popejoy, K.R. Crooks, S.M. Fullerton, L.A. Hindorff, G.W. Hooker, B.A. Koenig, N. Pino, E.M. Ramos, D.I. Ritter, H. Wand, M.W. Wright, M. Yudell, J.Y. Zou, S.E. Plon, C.D. Bustamante, K.E. Ormond, and the Clinical Genome Resource (ClinGen) Ancestry and Diversity Working Group, "Clinical Genetics Lacks Standard Definitions and Protocols for the Collection and Use of Diversity Measures," *American Journal of Human Genetics* 107, no. 1 (2020): 72–82.

13. Wendy D. Roth, "The Multiple Dimensions of Race," *Ethnic and Racial Studies* 39, no. 8 (2016): 1310–38.

14. Blair W. Weikel, Susanne Klawetter, Stephanie L. Bourque, Kathleen E. Hannan, Kristi Roybal, Modi Soondarotok, Marie St. Pierre, Yarden S. Fraiman, and Sunah S. Hwang, "Defining an Infant's Race and Ethnicity: A Systematic Review," *Pediatrics* 151, no. 1 (January 1, 2023): e2022058756. Birth certification did not become universal in the United States until 1933. Susan J. Pearson, *The Birth Certificate: An American History* (Chapel Hill: University of North Carolina Press, 2022).

15. Weikel et al., "Defining an Infant's Race and Ethnicity."

16. "How Is Race Determined on Birth Certificates?," US Birth Certificates, accessed February 9, 2024, https://www.usbirthcertificates.com/articles/race-on-birth-certificates.

17. Stacey A.N. D'Mello, Graeme J. Finlay, Bruce C. Baguley, and Marjan E. Askarian-Amiri, "Signaling Pathways in Melanogenesis," *International Journal of Molecular Sciences* 17, no. 7 (July 15, 2016): 1144.

18. M.O. Visscher, S.A. Burkes, D.M. Adams, A.M. Hammill, and R.R. Wickett, "Infant Skin Maturation: Preliminary Outcomes for Color and Biomechanical Properties," *Skin Research Technology* 23, no. 4 (November 2017): 545–51.

19. "How Is Race Determined on Birth Certificates?"

20. For other unintended consequences, see Steven Epstein, *Inclusion: The Politics of Difference in Medical Research* (Chicago: University of Chicago Press, 2007).

21. Weikel et al., "Defining an Infant's Race and Ethnicity."

22. Vickie M. Mays, Ninez A. Ponce, Donna L. Washington, and Susan D. Cochran, "Classification of Race and Ethnicity: Implications for Public Health," *Annual Review of Public Health* 24, no. 1 (2003): 83–110.

Conclusion

Clinical Implications

Meaghan O'Keefe and Cherie Ginwalla

On a sunny spring day, a hopeful couple goes through the paperwork they need to complete for the adoption agency they are working with. As they fill out the forms, they imagine the child that they will soon welcome into their family. They feel some fear but so much hope. What will the child look like? What characteristics did their birth parents have? Will the child look like the couple? Will they have the same skin tone, eye color, hair color and texture? Will others recognize their adopted child as theirs without question? How will their race affect their experience in the world? How may their health be affected by their race? What medical problems may their child face in the future?

Let us imagine some clinical encounters in this child's life. In the first, the newborn baby (like all babies) is screened for cystic fibrosis (CF) and sickle cell disease (SCD). In the second, what if this child has the misfortune to be diagnosed with a serious genetic disease? How might their perceived race or ethnicity affect how they are treated and what treatments are available? Third, imagine there is a new genetic therapy for the disease the child has. What barriers might the child face in accessing this treatment? Will such advances exacerbate inequity, or will they allow for more targeted interventions?

DIFFERENCES IN SCREENING

If this child is born in the present time in the United States, they will be screened for SCD and CF shortly after birth, regardless of their ethnicity. This was not always the case. For SCD, a national recommendation for universal newborn screening was made in 1987, but newborn screening did not become standard for all until 2006.[1] Universal newborn screening for CF was implemented in 2009. Both these diseases are autosomal recessive single-gene disorders (meaning that one would have to receive a disease-causing variant of the relevant gene from each parent in order to develop the disease) with serious health consequences. There are, however, substantial differences in testing and treatment and in the funding for research on these diseases. This is in part because these diseases have historically been racially coded. SCD is more common in people with African ancestry, and CF is more common in those of northern European and Ashkenazi Jewish ancestry.

In the medical field, race has often been used as a proxy for genetic traits or variants.[2] One particularly clear example of this is the way in which SCD has been identified and treated in the United States. SCD is an autosomal recessive single-gene disorder of the red blood cells that causes them to change their shape under stress. SCD is characterized by recurrent vaso-occlusive crises, which occur when the deformed red blood cells block blood vessels, causing excruciating pain and damaging vital organs, which can eventually reduce life expectancy. This genetic trait is found in people from geographic areas where malarial disease is endemic. Since most African Americans have ancestors who were brought to the United States from parts of Africa where malaria is endemic, it was thought to be a disease of African Americans through most of the twentieth century. When a blood test was developed that could detect sickle hemoglobin in the 1960s,[3] only African Americans were screened for the disease. SCD, however, is also prevalent in people of Mediterranean, Middle Eastern, and Indian descent. Indeed, there is a village in Greece in which 1 in 5 people have the disease.[4] In comparison, the rate for African Americans is 1 in 365.

When SCD screening programs for African Americans were introduced in the 1960s, the test identified people who had the disease, but it also identified those who carried the trait, meaning that they had inherited the disease-causing variant from only one parent. Carriers are far more common than those with the disease; among African Americans, between 7 and 9 people per 100 carry the trait.[5] At the start, African American communities supported these testing programs, believing that they would allow people to make informed decisions about reproduction and health.[6] By the 1970s, however, these tests had become a tool of discrimination.[7] Fourteen states made the tests a condition for accessing public education and for getting married. For many people, a positive carrier test resulted in higher insurance costs, job discrimination, and job loss.[8] In this case, we see how racially targeted disease screening for SCD failed to identify non–African American carriers and resulted in what the legal scholar Dorothy Roberts has rightly described as a disaster for the people who participated in testing.

As we have discussed elsewhere in this volume, race is a *product* of social processes that, in turn, structures modes of understanding that *produce* social and political effects.[9] In the case of SCD, we can see how a genetic trait was weaponized as a tool of control and discrimination. If we compare SCD to other genetic diseases, the differences in policies become very clear. For example, Tay-Sachs disease is, like SCD, an autosomal recessive single-gene mutation. A screening test for carriers was introduced in 1971,[10] around the same time as the SCD screening test. Genetic screening for Tay-Sachs was well received and popular in Ashkenazi Jewish communities (where 1 in 30 people is a carrier), and carriers experienced none of the punitive measures that marred SCD testing. For CF, screening tests were introduced in 1989. While this was later than many of the discriminatory practices outlined above, it is unlikely that a disease associated with northern Europeans would have resulted in anything like what happened with SCD.

Depending on the state the child is born in, they might first be given a test that measures blood levels of trypsinogen (a precursor to the enzyme trypsin) to screen for CF. If this precursor is elevated, they will be given a genetic test for CF. The child's genetic results will likely be evaluated using the CFTR2 database, which contains 159 variants that are implicated in CF. The database is fairly homogeneous in term of ancestral lineage—95 percent of the people who contributed are of European ancestry, which results in a test that is less likely to detect disease-causing variants that are more prevalent in people with non-European ancestry, such as the girl described in the chapter by Tina Rulli.[11] Moreover, different variants of CF have slightly different clinical indications. This means that symptoms from less common variants may not be recognized as being associated with CF because doctors tend to look for the symptoms of variants that are more common in European populations. A larger possible consequence is that future treatments will be designed to address symptoms more common in those of European ancestry and, thus, may be less effective in alleviating symptoms for patients with different variants.

DIFFERENCES IN TREATMENT

If the couple's child has CF or SCD, what kind of medical care might they expect? Both diseases are associated with complications that affect the quality of life and lead to a shortened lifespan for patients. Children with CF or SCD are frequently admitted to the hospital for more aggressive medical care than they can get at home. Patients with CF are frequently hospitalized for breathing difficulties and recurrent infections. The complications associated with CF include sinusitis, diabetes, pancreatic insufficiency resulting in difficulty with weight gain, growth and vitamin deficiencies, abdominal pain, liver dysfunction, bowel obstruction (distal intestinal obstruction syndrome, or DIOS), infertility, and others. Patients with SCD are hospitalized with painful crises and infections, often starting in infancy. They suffer from chronic pain throughout their lives. They also suffer from damage to many of their organs secondary to the sickling of their red blood cells and blockage of the blood vessels that supply oxygen to all the organs of the body.

Complications of SCD include strokes, recurrent infections, avascular necrosis of bone (death of bone cells), blood clots, kidney disease, vision loss, and others.

People of African or Indigenous American ancestry often face significant inequities in the management of their pain. To illustrate this, a study of children evaluated in the emergency department for abdominal pain found no racial differences in the testing done to evaluate the source of the abdominal pain—a symptom of both CF and SCD—but determined that significantly less pain medicine (particularly opioids) was administered to Black and Hispanic children compared with non-Hispanic white children. This is not just a matter of older physicians being trained at an earlier time, when racism may have been more common in medical practice. A study of misconceptions among medical trainees regarding biological differences between Black and white patients demonstrated that 25 percent of residents believed Black skin is thicker than white skin. Those who held these false beliefs were more likely to show bias in how much pain they perceived white people with the same condition experienced compared to Black people and thus were more likely to undertreat pain in Black patients.[12]

DIFFERENCES IN RESEARCH SUPPORT

Two early drug therapies—Pulmozyne, which thins mucus in the lungs, and TOBI, an aerosolized antibiotic specifically for CF—were the result of intentional directed research funded by the Cystic Fibrosis Foundation.[13] In 2019, the Food and Drug Administration approved gene modulation (cystic fibrosis transmembrane conductance regulator, or CFTR) therapy for CF for patients older than 12 years of age.[14] This was later expanded to include children over 2 years of age. The use of these therapies for this population of patients has been found to improve their lung function and decrease their need for hospitalizations. Treatment for CF has improved through targeted therapies, but these treatments are not effective in all patients, and there is still no cure, only some alleviation of symptoms.

While CF has benefited from sustained, directed, well-funded research, there are disparities in research funding when it comes to SCD. SCD is three times as common as CF, but the two diseases have received the same amount in federal funding. When private funding is factored in, the disparity increases exponentially. For example, in the period from 2013 to 2016, CF research received 971 times more funding than SCD.[15] This discrepancy has resulted in fewer research articles and fewer drug approvals for SCD. There have also been innovations in treatment for SCD, but these discoveries have been more accidental than intentional.[16] For example, one of the most common medications prescribed to relieve symptoms, hydroxyurea, was initially used in chemotherapy, but starting in the 1980s, physicians began using it for SCD. It took until 1998 for hydroxyurea to be approved by the FDA for adults with SCD. The only curative treatment was discovered when a patient with leukemia was given a hematopoietic stem cell transplant, which also cured their SCD. While these transplants do cure SCD, it is difficult to find matching donors, and there is a serious risk of adverse effects.[17]

TECHNOLOGICAL PROMISE AND ECONOMIC ACCESS

More recently, however, gene therapies targeting variants that cause CF and SCD have been developed. These biotechnologies have the potential to radically improve outcomes, especially for those with SCD. As of November 2023, the FDA has indicated that it will likely approve a gene therapy for SCD developed using the CRISPR technology, the first approval of a therapy that uses the new genetic medication technique.[18] The treatment works by removing the bone marrow cells and modifying the gene that governs the production of red blood cells. The patient then undergoes chemotherapy to eradicate the remaining cells that still have the genetic mutation for SCD.[19] The cells with the modified genes are then infused back into the patient's body. This treatment does not work by fixing the mutation; rather, it uses a compensatory mechanism to stimulate the production of fetal hemoglobin cells.[20] These new cells are able to carry oxygen through the body more effectively than sickled cells.[21] Although there are currently no similar gene therapies for CF, research is underway to develop them.

If the couple adopting this child has good health insurance and a relatively high income, they will most likely be able to benefit from these innovations. If not, these new interventions may be out of reach. In our largely for-profit medical system, novel and expensive treatments tend to exacerbate existing health inequalities. Gene therapies similar to those that may soon be available for SCD carry a price tag of up to $3.5 million per patient; researchers have speculated that the price for the new SCD therapy will be between $4 million and $6 million.[22] This price will go down as the technology advances, but it is unlikely to become anything close to affordable in the foreseeable future. This will likely be true of future genetic therapies for CF, although racialized health insurance and wealth gaps[23] will affect availability: studies have shown that extremely expensive medical treatments like this one are hard for people to access because of "discriminatory insurance coverage, onerous reimbursement payee issues, and severe copay burdens."[24] For now, even if the cost is somehow mitigated, these therapies will be available only at major medical research facilities, which also makes them geographically difficult to access for most people.

GENES AND ENVIRONMENT

Throughout this conclusion, we have been discussing single-gene mutation diseases, which follow a clear, Mendelian inheritance process. Most diseases do not adhere to such a clear pattern. While there may be genetic components to susceptibility to diseases and the kinds of symptoms and outcomes people experience, there is considerable evidence that environment and health access are more determinative in the context of health. Take, for example, asthma, a complex disease affecting 8 percent of American children,[25] which seems to have a hereditary component but is also substantially influenced by a child's living environment. For example, as a group, Puerto Ricans have one of the highest rates of asthma at 14.9 percent.[26] Puerto Ricans also have ancestry lineages that are quite distant from one another. Even though there is a range in terms of ancestry-informative genetic markers,[27] researchers have sought evidence of a founder effect, which would mean that Puerto Ricans are descended from a limited number of individuals, one or more of whom had asthma-causing variants.[28] Medical geneticists have suggested that studying communities with these kinds of founder effects might be an effective tool in identifying patients at higher risk for diseases that might not be identifiable with standard population groups.[29] In other words, given that Puerto Ricans have diverse ancestral lineages but are descended from a smaller number of people, researchers expect that using a Puerto Rican dataset to examine disease risk would be far more effective in identifying the variants associated with asthma than using existing datasets that use broader population labels.

Undoubtedly, the availability of more diverse and more fine-scaled genetic databases, combined with standardized electronic health records, could benefit a broad range of people, allowing them to take health precautions or start early treatment of diseases. There is a risk, however, that when diseases are coded as genetic, then researchers and practitioners give less attention to nongenetic factors. Even though asthma seems to have a genetic component, environmental factors—such as living in a household with a smoker, air pollution, and allergens—play an enormous role in the development of disease.[30] In fact, when researchers compared Puerto Ricans living in New York with Puerto Ricans living in Puerto Rico, where fewer people smoke, the air is less polluted, and allergens linked to asthma are less common, they found far lower rates, between 6.4 and 7.7 percent[31] as opposed to 14.9 percent. In such cases, attributing the asthma to genetics can mask the effects of environment and, importantly, the effects of racial disparities in health-care access and access to healthy living conditions.[32]

We began this volume with the process of selecting gametes and embryos and how these decisions are shaped not simply by the hopes and desires of the prospective parents but also by systems, institutions, and practices already in place. Many of the chapters addressed the ways in which people are racialized and the ways racialization is reproduced socially and scientifically. In a racialized society, we are all assigned a racial identity—a process in which, as the chapter by Alice B. Popejoy shows, we may or may not have much agency—and those identities structure our relationships, opportunities, and experiences. These racialized identities also affect how patients are tested and diagnosed, how they are treated by health professionals, and ultimately how funding is allocated for biomedical research. Given this history, we suggest that future medical research ought to adopt fine-grained genetic analysis based on a continuous model of ancestry, rather than one using continental clusters. More importantly, we also hope that any genetic approach to medical research or practice takes seriously the effects of social inequalities and racial discrimination.

NOTES

1. Jane M. Benson and Bradford L. Therrell Jr., "History and Current Status of Newborn Screening for Hemoglobinopathies," *Seminars in Perinatology* 34, no. 2 (2010): 134–44.

2. Michael Root, "The Use of Race in Medicine as a Proxy for Genetic Differences," *Philosophy of Science* 70, no. 5 (2003): 1173–83.

3. Dorothy Roberts, *Killing the Black Body: Race, Reproduction, and the Meaning of Liberty* (New York: Vintage, 1997), 254.

4. Troy Duster, Pooya Naderi, and Sonia Rosales, "Interview with Troy Duster, New York University," *Social Thought and Research* 29 (2008): 31–47, 31.

5. Vimal K. Derebail, Patrick H. Nachman, Nigel S. Key, Heather Ansede, Ronald J. Falk, and Abhijit V. Kshirsagar, "High Prevalence of Sickle Cell Trait in African Americans with ESRD," *Journal of the American Society of Nephrology: JASN* 21, no. 3 (2010): 413–17.

6. Roberts, *Killing the Black Body*, 255.

7. See Roberts, *Killing the Black Body*, 255–57; and Troy Duster, *Backdoor to Eugenics*, 2nd ed. (New York: Routledge, 2003), 47–50.

8. Roberts, *Killing the Black Body*, 256.

9. Camisha A. Russell, *The Assisted Reproduction of Race* (Bloomington: Indiana University Press, 2018), 2.

10. Michael Kaback, Joyce Lim-Steele, Deepti Dabholkar, David Brown, Nancy Levy, and Karen Zeiger, "Tay-Sachs Disease—Carrier Screening, Prenatal Diagnosis, and the Molecular Era: An International Perspective, 1970 to 1993," *JAMA* 270, no. 19 (1993): 2307–15.

11. Regine M. Lim, Ari J. Silver, Maxwell J. Silver, Carlos Borroto, Brett Spurrier, Tanya C. Petrossian, Jessica L. Larson, and Lee M. Silver, "Targeted Mutation Screening Panels Expose Systematic Population Bias in Detection of Cystic Fibrosis Risk," *Genetics in Medicine* 18, no. 2 (2016): 174–79.

12. Kelly M. Hoffman, Sophie Trawalter, Jordan R. Axt, and M. Norman Oliver, "Racial Bias in Pain Assessment and Treatment Recommendations, and False Beliefs about Biological Differences between Blacks and Whites," *Proceedings of the National Academy of Sciences* 113, no. 16 (2016): 4296–301.

13. "Our History," Cystic Fibrosis Foundation, accessed February 9, 2024, https://cff.org/about-us /our-history.

14. M. S. Schechter, N. Sabater-Anaya, G. Oster, D. Weycker, H. Wu, E. Arteaga-Solis, S. Bagal, L. J. McGarry, K. Van Brunt, and J. M. Geiger, "Impact of Elexacaftor/Tezacaftor/Ivacaftor on Healthcare Resource Utilization and Associated Costs among People with Cystic Fibrosis in the U.S.: A Retrospective Claims Analysis," *Pulmonary Therapy* 9, no. 4 (2023): 479–98.

15. Faheem Farooq and John J. Strouse, "Disparities in Foundation and Federal Support and Development of New Therapeutics for Sickle Cell Disease and Cystic Fibrosis," *Blood* 132, Supplement 1 (2018): 4687.

16. Farooq and Strouse, "Disparities in Foundation and Federal Support," 7.

17. Farooq and Strouse, "Disparities in Foundation and Federal Support," 7.

18. Rob Stein, "FDA Advisers See No Roadblock for Gene-Editing Treatment for Sickle Cell Disease," NPR, October 31, 2023, https://npr.org/sections/health-shots/2023/10/31/1208041252/a-landmark -gene-editing-treatment-for-sickle-cell-disease-moves-closer-to-reality.

19. Stein, "FDA Advisers See No Roadblock."

20. Cormac Sheridan, "The World's First CRISPR Therapy Is Approved: Who Will Receive It?," *Nature Biotechnology* 42, no. 1 (2024): 3–4.

21. Sheridan, "World's First CRISPR Therapy."

22. Andrew M. Subica, "CRISPR in Public Health: The Health Equity Implications and Role of Community in Gene-Editing Research and Applications," *American Journal of Public Health* 113, no. 8 (2023): 874–82.

23. Courtney Boen, Lisa Keister, and Brian Aronson, "Beyond Net Worth: Racial Differences in Wealth Portfolios and Black–White Health Inequality across the Life Course," *Journal of Health and Social Behavior* 61, no. 2 (2020): 153–69.

24. Subica, "CRISPR in Public Health," 877.

25. "Asthma Surveillance Data," Centers for Disease Control and Prevention, March 29, 2023, https://cdc.gov/asthma/asthmadata.htm.

26. "Asthma and Hispanic Americans," US Department of Health and Human Services, Office of Minority Health, accessed November 10, 2023, https://minorityhealth.hhs.gov/asthma-and-hispanic -americans.

27. Sylvia E. Szentpetery, Erick Forno, Glorisa Canino, and Juan C. Celedón, "Asthma in Puerto Ricans: Lessons from a High-Risk Population," *Journal of Allergy and Clinical Immunology* 138, no. 6 (2016): 1556–58.

28. Gillian M. Belbin, Sinead Cullina, Stephane Wenric, Emily R. Soper, Benjamin S. Glicksberg, Denis Torre, Arden Moscati, et al., "Toward a Fine-Scale Population Health Monitoring System," *Cell* 184, no. 8 (2021): 2068–83.

29. Belbin et al., "Toward a Fine-Scale Population Health Monitoring System," 2078.

30. Suad El Burai Félix, Cathy M. Bailey, and Hatice S. Zahran, "Asthma Prevalence among Hispanic Adults in Puerto Rico and Hispanic Adults of Puerto Rican Descent in the United States—Results from Two National Surveys," *Journal of Asthma* 52, no. 1 (2015): 3–9.

31. Félix et al., "Asthma Prevalence among Hispanic Adults."

32. Anna C. F. Lewis, Santiago J. Molina, Paul S. Appelbaum, Bege Dauda, Anna Di Rienzo, Agustin Fuentes, Stephanie M. Fullerton, et al., "Getting Genetic Ancestry Right for Science and Society," *Science* 376, no. 6590 (2022): 250–52.

BIBLIOGRAPHY

Adigbli, George. "Race, Science and (Im)precision Medicine." *Nature Medicine* 26 (2020): 1675–76.

Adler, Simon. "G: Unnatural Selection." *RadioLab*, July 25, 2019, https://www.nycstudios .org/podcasts/radiolab/articles/g-unnatural-selection.

Almeling, Rene. *Sex Cells: The Medical Market for Eggs and Sperm.* Berkeley: University of California Press, 2011.

Ancestry. https://ancestry.com.

Ancestry. "Your DNA Results, with More Detail than Ever." https://ancestry.com/c/dna /ancestry-dna-ethnicity-estimate-update.

AncestryDNA. https://ancestry.com/dna.

Anderson, Karen S. "Massive Resistance, Violence, and Southern Social Relations: The Little Rock, Arkansas, School Integration Crisis, 1954–1960." In *Massive Resistance: Southern Opposition to the Second Reconstruction*, ed. Clive Webb, 66–72. New York: Oxford University Press, 2005.

Appiah, Kwame Anthony. "Reconstructing Racial Identities." *Research in African Literatures* 27 (1996): 68–72.

Asbury, Kathryn, and Robert Plomin. *G is for Gene: The Impact of Genetics on Education and Achievement.* London: John Wiley and Sons, 2013.

Balmer, Randall. *Bad Faith: Race and the Rise of the Religious Right.* Grand Rapids, MI: Eerdmans, 2021.

Balter, Michael. "New Mystery for Native American Origins." *Science* 349, no. 6246 (2015): 354–55.

Barcellos, Sylvia H., Leandro Carvalho, and Patrick Turley. "The Effect of Education on the Relationship between Genetics, Early-Life Disadvantages, and Later-Life SES." *National Bureau of Economic Research* w28750 (2021). https://www.nber.org/papers /w28750.

Barkan, Elazar. *The Retreat of Scientific Racism: Changing Concepts of Race in Britain and the United States between the World Wars.* New York: Cambridge University Press, 1992.

Barragán, Carlos Andrés. "Lineages within Genomes: Situating Human Genetics Research and Contentious Bio-Identities in Northern South America." PhD diss., University of California, Davis, 2016.

——. "Molecular Vignettes of the Colombian Nation: The Place(s) of Race and Ethnicity in Networks of Biocapital." In *Racial Identities, Genetic Ancestry and Health in South America*, ed. Sahra Gibbon, Ricardo Ventura Santos, and Mónica Sans, 41–68. New York: Palgrave Macmillan, 2012.

——. "Untangling Population Mixture? Genomic Admixture and the Idea of *Mestizos* in Latin America." *Gene Watch* 28, no. 2 (2015): 11–13, 20–21.

Bashford, Alison, and Philippa Levine, eds. *The Oxford Handbook of the History of Eugenics.* New York: Oxford University Press, 2010.

BBC. "Nivea Removes Its 'White is Purity' Deodorant Advert Branded 'Racist.'" April 4, 2017. https://bbc.com/news/world-europe-39489967.

Belbin, Gillian M., Sinead Cullina, Stephane Wenric, Emily R. Soper, Benjamin S. Glicksberg, Denis Torre, Arden Moscati, et al. "Toward a Fine-Scale Population Health Monitoring System." *Cell* 184, no. 8 (2021): 2068–83.

Bell, Duncan. *Dreamworlds of Race: Empire and the Utopian Destiny of Anglo-America.* Princeton, NJ: Princeton University Press, 2020.

Bennett, Matthew R., David Bustos, Jeffrey S. Pigati, Kathleen B. Springe, Thomas M. Urban, Vance T. Holliday, Sally C. Reynolds, Marcin Budka, Jeffrey S. Honke, Adam M. Hudson, Brendan Fenerty, Clare Connelly, Patrick J. Martinez, Vincent L. Santucci, and Daniel Odess. "Evidence of Humans in North America during the Last Glacial Maximum." *Science* 373, no. 6562 (2021): 1528–31.

Benson, Jane M., and Bradford L. Therrell Jr. "History and Current Status of Newborn Screening for Hemoglobinopathies." *Seminars in Perinatology* 34, no. 2 (2010): 134–44.

Birenbaum-Carmeli, Daphna, Yoram Carmeli, and Sergei Gornostayev. "Researching Sensitive Fields: Some Lessons from a Study of Sperm Donors in Israel." *International Journal of Sociology and Social Policy* 28, no. 11/12 (2008): 425–39.

Blau, Peter M, and Otis Dudley Duncan. *The American Occupational Structure.* Hoboken, NJ: John Wiley and Sons, 1967.

Bliss, Catherine. "Mapping Race through Admixture." *International Journal of Technology, Knowledge and Society* 4, no. 4 (2008): 79–83.

——. *Race Decoded: The Genomic Fight for Social Justice.* Stanford, CA: Stanford University Press, 2012.

Block, Ned. "How Heritability Misleads about Race." *Cognition* 56 (1995): 99–128.

Boardman, Jason D., and Jason M. Fletcher. "Evaluating the Continued Integration of Genetics into Medical Sociology." *Journal of Health and Social Behavior* 62, no. 3 (2021): 401–81.

Bodmer, Walter F., and Luigi Luca Cavalli-Sforza. "Intelligence and Race." *Scientific American*, October 1, 1970. https://scientificamerican.com/article/intelligence-and-race.

Boen, Courtney, Lisa Keister, and Brian Aronson. "Beyond Net Worth: Racial Differences in Wealth Portfolios and Black-White Health Inequality across the Life Course." *Journal of Health and Social Behavior* 61, no. 2 (2020): 153–69.

Bolnick, Deborah A. "Individual Ancestry Inference and the Reification of Race as a Biological Phenomenon." In *Revisiting Race in a Genomic Age*, ed. Barbara A. Koenig,

Sandra Soo-Jin Lee, and Sarah S. Richardson, 70–85. New Brunswick, NJ: Rutgers University Press, 2008.

Bolnick, Deborah A., Duana Fullwiley, Troy Duster, Richard S. Cooper, Joan H. Fujimura, Jonathan Kahn, Jay S. Kaufman, Jonathan Marks, Ann Morning, Alondra Nelson, Pilar Ossorio, Jenny Reardon, Susan M. Reverby, and Kimberly TallBear. "The Science and Business of Genetic Ancestry Testing." *Science* 318 (2007): 399–400.

Bonham, Vence L., Sherrill L. Sellers, Thomas H. Gallagher, Danielle Frank, Adebola O. Odunlami, Eboni G. Price, and Lisa A. Cooper. "Physicians' Attitudes toward Race, Genetics, and Clinical Medicine." *Genetics in Medicine* 11, no. 4 (2009): 279–86.

Bonilla-Silva, Eduardo. *Racism without Racists: Color-Blind Racism and the Persistence of Racial Inequality in the United States.* Lanham, MD: Rowman and Littlefield, 2006.

Bowker, Geoffrey C., and Susan Leigh Star. *Sorting Things Out: Classification and Its Consequences.* Cambridge, MA: MIT Press, 2000.

Bowles, Nellie. "The Sperm Kings Have a Problem: Too Much Demand." *New York Times,* January 8, 2021. https://www.nytimes.com/2021/01/08/business/sperm-donors-facebook -groups.html.

Boyarin, Jonathan. *The Unconverted Self: Jews, Indians, and the Identity of Christian Europe.* Chicago: University of Chicago Press, 2009.

Boyd, R. "Kinds as the 'Workmanship of Men': Realism, Constructivism, and Natural Kinds." In *Rationalität, Realismus, Revision,* ed. J. Nida-Rümelin, 52–89. Internationalen Kongresses des Gesellschaft für Analytische Philosophie, 1999.

———. "Rethinking Natural Kinds, Reference and Truth: Towards More Correspondence with Reality, Not Less." *Synthese* 198 (2021): 2863–903.

Branson, Susan. "Phrenology and the Science of Race in Antebellum America." *Early American Studies: An Interdisciplinary Journal* 15, no. 1 (2017): 164–93.

Braude, Benjamin. "The Sons of Noah and the Construction of Ethnic and Geographical Identities in the Medieval and Early Modern Periods." *William and Mary Quarterly* 54, no. 1 (1997): 103–42.

Braun, Lundy. "Race, Ethnicity, and Lung Function: A Brief History." *Canadian Journal of Respiratory Therapy* 51, no. 4 (2015): 99–101.

Brigham, Carl Campbell. "Intelligence Tests of Immigrant Groups." *Psychological Review* 37, no. 3 (1930): 158–65.

Bromberg, P.M. "Shadow and Substance: A Relational Perspective on Clinical Process." *Psychoanalytic Psychology* 10 (1993): 147–68.

Brooks, D.S., J. DiFrisco, and W.C. Wimsatt. *Levels of Organization in the Biological Sciences.* Cambridge, MA: MIT Press, 2021.

Brown, Anna. "The Changing Categories the U.S. Census Has Used to Measure Race." Pew Research Center, 2020. https://www.pewresearch.org/fact-tank/2020/02/25/the-changing -categories-the-u-s-has-used-to-measure-race.

Bueno, O. "Structural Realism, Mathematics, and Ontology." *Studies in History and Philosophy of Science* 74 (2019): 4–9.

Burge, Ryan P., and Miles D. Williams. "Is Social Media a Digital Pulpit? How Evangelical Leaders Use Twitter to Encourage the Faithful and Publicize Their Work." *Journal of Religion, Media and Digital Culture* 8, no. 3 (2019): 309–39.

Burks, Barbara. "The Relative Influence of Nature and Nurture upon Mental Development: A Comparative Study of Foster Parent-Foster Child Resemblance and True Parent-True

Child Resemblance." *Twenty-Seventh Yearbook of the National Society for the Study of Education, Part 1* (1928): 219–316.

Burt, Callie. "Challenging the Utility of Polygenic Scores for Social Science: Environmental Confounding, Downward Causation, and Unknown Biology." *Behavioral and Brain Sciences* 46 (2023): e207.

Business Wire. "U.S. Fertility Clinics Market Report 2023." July 13, 2023. https://businesswire.com/news/home/20230713777238/en/US-Fertility-Clinics-Market-Report-2023-Sector-is-Expected-to-Reach-16.8-Billion-by-2028-at-a-CAGR-of-13.6.

Bustamante, Carlos D., Esteban Gonzalez Burchard, and Francisco M. De La Vega. "Genomics for the World." *Nature* 475, no. 7355 (2011): 163–65.

Butler, Anthea. *White Evangelical Racism: The Politics of Morality in America.* Chapel Hill: University of North Carolina Press, 2021.

California Cryobank. https://cryobank.com.

California Cryobank. "Choosing Your Donor." https://cryobank.com/how-it-works/choosing-your-donor.

———. "DNA Ancestry." https://cryobank.com/services/dna-ancestry.

———. "DNA Ancestry Page Sample Information." https://cryobank.com/_resources/pdf/sampleinformation/dnaancestrysample.pdf.

———. "Donor Information." https://cryobank.com/donor-search/donor-information/#Donor-Profiles.

———. "Donor Qualification." https://cryobank.com/how-it-works/donor-qualification.

———. "Donor Recruitment." https://cryobank.com/how-it-works/donor-recruitment.

———. "Donor Search." https://cryobank.com/search.

Callaway, Ewen. "Migration to Americas Traced." *Nature* 563, no. 7731 (2018): 303–4.

Carson, John S. "Army Alpha, Army Brass, and the Search for Army Intelligence." *Isis* 84, no. 2 (1993): 278–309.

———. *The Measure of Merit: Talents, Intelligence, and Inequality in the French and American Republics, 1750–1940.* Princeton, NJ: Princeton University Press, 2007.

Carter, J. E. L., and B. H. Heath. "Somatotype Methodology and Kinesiology Research." *Kinesiology Review* 10 (1971): 10–19.

Carter, J. Kameron. *Race: A Theological Account.* New York: Oxford University Press, 2008.

Caspi, Avshalom, Joseph McClay, Terrie E. Moffitt, Jonathan Mill, Judy Martin, Ian W. Craig, Alan Taylor, and Richie Poulton. "Role of Genotype in the Cycle of Violence in Maltreated Children." *Science* 297 (2002): 851–54.

Cavalli-Sforza, Luigi Luca. "Human Genome Diversity: Where Is the Project Now?" In *Human DNA: Law and Policy*, ed. Bartha Maria Knoppers, Claude M. Laberge, and Marie Hirtle, 219–27. Boston: Kluwer Law International, 1997.

Cavalli-Sforza, Luigi Luca, Menozzi Piazza, and Alberto Piazza. *The History and Geography of Human Genes.* Princeton, NJ: Princeton University Press, 1994.

Cavalli-Sforza, Luigi Luca, Allan C. Wilson, Charles R. Cantor, Robert M. Cook-Deegan, and Mary-Claire King. "Call for a Worldwide Survey of Human Genetic Diversity: A Vanishing Opportunity for the Human Genome Project." *Genomics* 11, no. 2 (1991): 490–91.

Centers for Disease Control and Prevention. "Asthma Surveillance Data." March 29, 2023. https://cdc/gov/asthma/asthmadata.htm.

———. "State-Specific Assisted Reproductive Technology Surveillance." https://archive.cdc .gov/#/details?url=https://www.cdc.gov/art/state-specific-surveillance/index.html.

Cerdeña, Jessica P., Emmanuella Ngozi Asabor, Marie V. Plaisime, and Rachel R. Hardeman. "Race-Based Medicine in the Point-of-Care Clinical Resource UpToDate: A Systematic Content Analysis." *eClinicalMedicine* 52 (2022). https://doi.org/10.1016/j .eclinm.2022.101581.

Cerdeña, Jessica P., Vanessa Grubbs, and Amy L. Non. "Genomic Supremacy: The Harm of Conflating Genetic Ancestry and Race." *Human Genomics* 16 (2022): 18.

Chabris, Christopher F., B. M. Hebert, Daniel J. Benjamin, Jonathan P. Beauchamp, David Cesarini, M. van der Loos, M. Johannesson, P. K. Magnusson, P. Lichtenstein, C. S. Atwood, Jeremy Freese, T. S. Hauser, Robert M. Hauser, Nicholas Christakis, and David I. Laibson. "Most Reported Genetic Associations with General Intelligence Are Probably False Positives." *Psychological Science* 23, no. 11 (2012): 1314–23.

Chabris, Christopher F., James J. Lee, Daniel J. Benjamin, Jonathan P. Beauchamp, Edward L. Glaeser, Gregoire Borst, Steven Pinker, and David I. Laibson. "Why It Is Hard to Find Genes Associated with Social Science Traits: Theoretical and Empirical Considerations." *American Journal of Public Health* 103, supplement 1 (2013): S152–66.

Chabris, Christopher F., James J. Lee, David Cesarini, Daniel J. Benjamin, and David I. Laibson. "The Fourth Law of Behavior Genetics." *Current Directions in Psychological Science* 24, no. 4 (2015): 304–12.

Chakraborty, Ranajit, and Kenneth M. Weiss. "Admixture as a Tool for Finding Linked Genes and Detecting That Difference from Allelic Associations between Loci." *Proceedings of the National Academy of Sciences* 85, no. 23 (1988): 9119–23.

Cheesman, Rosa, Avina Hunan, Jonathan R. I. Coleman, Yasmin Ahmadzadeh, Robert Plomin, Tom A. McAdams, Thalia C. Eley, and Gerome Breen. "Comparison of Adopted and Non-Adopted Individuals Reveals Gene-Environment Interplay for Education in the U.K. Biobank." *Psychological Science* 31, no. 5 (2020): 582–91.

Christopher, Gail C., Nikole Hannah-Jones, and Derrick Johnson. "Racial Hierarchy, Race Narrative, and the Structures that Sustain Them." In *Necessary Conversations: Understanding Racism as a Barrier to Achieving Health Equity*, ed. Alonzo L. Plough, 19–33. New York: Oxford University Press, 2022.

Clark, Gregory. "The Inheritance of Social Status: England 1600 to 2022." *Proceedings of the National Academy of Sciences* 120, no. 27 (2023): e2300926120.

Clarke, Adele E., Janet K. Shim, Laura Mamo, J. R. Fosket, and Jennifer R. Fishman. "Biomedicalization: Technoscientific Transformations of Health, Illness, and U.S. Biomedicine." *American Sociological Review* 68, no. 2 (2003): 161–94.

Coalition for Biblical Sexuality. "Nashville Statement." https://cbmw.org/nashville -statement.

Collins, Simone, and Malcolm Collins. "Why the World Needs More Big Families like Ours amid the Population Crisis." *New York Post*, January 28, 2023. https://nypost .com/2023/01/28/the-world-needs-more-big-families-like-ours-for-humans-to-survive.

Collins-Schramm, Heather E., Bill Chima, Takanobu Morii, Kimberly Wah, Yolanda Figueroa, Lindsey A. Criswell, Robert L. Hanson, William C. Knowlder, Gabriel Silva, John W. Belmont, and Michael F. Seldin. "Mexican American Ancestry-Informative Markers: Examination of Population Structure and Marker Characteristics in European

Americans, Mexican Americans, Amerindians and Asians." *Human Genetics* 114, no. 3 (2004): 263–71.

Comfort, Nathaniel. *The Science of Human Perfection: How Genes Became the Heart of American Medicine*. New Haven, CT: Yale University Press, 2012.

Conley, Dalton. "What if Tinder Showed Your IQ?" *Nautilus*, September 10, 2015. https://nautil.us/what-if-tinder-showed-your-iq-235618.

Conley, Dalton, and Jason M. Fletcher. *The Genome Factor: What the Social Genomics Revolution Reveals about Ourselves, Our History, and the Future*. Princeton, NJ: Princeton University Press, 2017.

Conley, Shannon N. "Who Gets to Be Born? The Anticipatory Governance of Pre-Implantation Genetic Diagnosis Technology in the United Kingdom from 1978–2001." *Journal of Responsible Innovation* 7, no. 3 (2020): 507–27.

Cook-Deegan, Robert Mullan. "Origins of the Human Genome Project." *RISK: Health, Safety & Environment* 5, no. 2 (1994): 97–118.

Coop, Graham. "Genetic Similarity versus Genetic Ancestry Groups as Sample Descriptors in Human Genetics." arXiv (2022). https://arxiv.org/pdf/2207.11595.pdf.

Coop, Graham, Michael B. Eisen, Rasmus Nielsen, Molly Przeworski, Noah Rosenberg, et al. Letter to the editor. *New York Times*, Book Review, August 18, 2014. https://nytimes.com/2014/08/10/books/review/letters-a-troublesome-inheritance.html.

Coop, Graham, and Molly Przeworski. "Lottery, Luck, or Legacy: A Review of *The Genomic Lottery: Why DNA Matters for Social Equality*." *Evolution* 76, no. 4 (2022): 846–53.

Cooper, R. M., and John P. Zubek. "Effects of Enriched and Restricted Early Environments on the Learning Ability of Bright and Dull Rats." *Canadian Journal of Psychology* 12, no. 3 (1958): 159–64.

Cowan, Ruth Schwartz. "Nature and Nurture: The Interplay of Biology and Politics in the Work of Francis Galton." *Studies in the History of Biology* 1 (1977): 133–208.

Cox, Daniel, and Robert P. Jones. "Deep Divide between Black and White Americans in Views of Criminal Justice System." Public Religion Research Institute, May 7, 2015. https://prri.org/research/divide-white-black-americans-criminal-justice-system.

Craver, C. F. "Mechanisms and Natural Kinds." *Philosophical Psychology* 22 (2009): 575–94.

Cromer, Risa. *Conceiving Christian America: Embryo Adoption and Reproductive Politics*. New York: NYU Press, 2023.

———. "Making the Ethnic Embryo: Enacting Race in U.S. Embryo Adoption." *Medical Anthropology* 38, no. 7 (2019): 603–19.

———. "'Our Family Picture Is a Little Hint of Heaven': Race, Religion and Selective Reproduction in U.S. 'Embryo Adoption.'" *Reproductive Biomedicine and Society Online* 11 (2020): 9–17.

Curtis, Jesse. *The Myth of Colorblind Christians: Evangelicals and White Supremacy in the Civil Rights Era*. New York: NYU Press, 2021.

Cystic Fibrosis Foundation. "Our History." https://cff.org/about-us/our-history.

Dailey, Jane. "Sex, Segregation, and the Sacred after *Brown*." *Journal of American History* 91, no. 1 (2004): 119–44.

Daniels, Cynthia R., and Janet Golden. "Procreative Compounds: Popular Eugenics, Artificial Insemination and the Rise of the American Sperm Banking Industry." *Journal of Social History* 38, no. 1 (2004): 5–27.

Daston, Lorraine, and Peter Galison. *Objectivity*. Princeton, NJ: Princeton University Press, 2007.

Davis, Dána-Ain. "Reproducing while Black: The Crisis of Black Maternal Health, Obstetric Racism and Assisted Reproductive Technology." *Reproductive Biomedicine & Society Online* 11 (2020): 56–64.

Davis, Leniece T. "Stranger in Mine Own House: Double-Consciousness and American Citizenship." In *Contemporary Patterns of Politics, Praxis, and Culture*, ed. Georgia A. Persons, 148–53. New York: Routledge, 2005.

Denbow, Jennifer M. *Governed through Choice: Autonomy, Technology, and the Politics of Reproduction*. New York: NYU Press, 2015.

Deomampo, Daisy. "Race, Nation, and the Production of Intimacy: Transnational Ova Donation in India." *Positions: East Asia Cultures Critique* 24, no. 1 (2016): 303–32.

———. "Racialized Commodities: Race and Value in Human Egg Donation." *Medical Anthropology* 38, no. 7 (2019): 620–33.

Derebail, Vimal K., Patrick H. Nachman, Nigel S. Key, Heather Ansede, Ronald J. Falk, and Abhijit V. Kshirsagar. "High Prevalence of Sickle Cell Trait in African Americans with ESRD." *Journal of the American Society of Nephrology: JASN* 21, no. 3 (2010): 413–17.

Devitt, James. "Genes Have Small Effect on Length of Education." *Futurity*, June 3, 2013. https://www.futurity.org/genes-have-small-effect-on-length-of-education.

De Vries, Jantina, Susan J. Bull, Ogobara Doumbo, Muntaser Ibrahim, Odile Mercereau-Puijalon, Dominic Dwiatkowski, and Michael Parker. "Ethical Issues in Human Genomics Research in Developing Countries." *BMC Medical Ethics* 12 (2011): 5.

D'Mello, Stacey A. N., Graeme J. Finlay, Bruce C. Baguley, and Marjan E. Askarian-Amiri. "Signaling Pathways in Melanogenesis." *International Journal of Molecular Sciences* 17, no. 7 (July 15, 2016): 1144.

Donnelly, John. "Comparative Effectiveness Research." *Health Affairs Health Policy Brief*, October 5, 2010.

DonorNexus. "Premier Egg Donors." https://donornexus.com/services/egg-donation/premier-egg-donor-cycle.

Dougherty, Kevin D., Mark Chaves, and Michael O. Emerson. "Racial Diversity in U.S. Congregations, 1998–2019." *Journal for the Scientific Study of Religion* 59, no. 4 (2020): 651–62.

Driscoll, Mark. "Church: What Are the Characteristics of the Church?" Real Faith by Mark Driscoll. https://realfaith.com/what-christians-believe/characteristics-church.

———. "What Does Depravity Mean and Not Mean?" Real Faith by Mark Driscoll. https://realfaith.com/daily-devotions/what-does-total-depravity-mean-and-not-mean.

Dudbridge, F. "Power and Predictive Accuracy of Polygenic Risk Scores." *PLoS Genetics* 9 (2013): e1003348.

Duello, Theresa M., Shawna Rivedal, Colton Wickland, and Annika Weller. "Race and Genetics vs. 'Race' in Genetics: A Systematic Review of the Use of African Ancestry in Genetic Studies." *Evolution, Medicine, and Public Health* 9, no. 1 (2021): 232–45.

Du Mez, Kristin Kobes. *Jesus and John Wayne: How White Evangelicals Corrupted a Faith and Fractured a Nation*. New York: Liveright, 2020.

Durand, Eric Y., Chuong B. Do, Peter R. Wilton, Joanna L. Mountain, Adam Auton, G. David Poznik, and J. Michael Macpherson. "A Scalable Pipeline for Local Ancestry

Inference Using Tens of Thousands of Reference Haplotypes." bioRxiv (2021). https://doi.org/10.1101/2021.01.19.427308.

Duster, Troy. *Backdoor to Eugenics*. 2nd ed. New York: Routledge, 2003.

Duster, Troy, Pooya Naderi, and Sonia Rosales. "Interview with Troy Duster, New York University." *Social Thought and Research* 29 (2008): 31–47.

Eaton, Jennifer L., Tracy Truong, Yi-Ju Li, and Alex Polotsky. "Prevalence of a Good Perinatal Outcome with Cryopreserved Compared with Fresh Donor Oocytes." *Obstetrics and Gynecology* 135, no. 3 (2020): 709–16.

Elisha, Omri. "Moral Ambitions of Grace: The Paradox of Compassion and Accountability in Evangelical Faith-Based Activism." *Cultural Anthropology* 23, no. 1 (2008): 154–89.

Emergen Research. "Market Synopsis: Direct-to-Consumer Genetic Testing Market, by Test Type (Carrier Testing and Predictive Testing), by Technology (Targeted Analysis and Whole Genome Sequencing), and by Region Forecast to 2030." September 2022. https://emergenresearch.com/industry-report/direct-to-consumer-genetic-testing-market.

Emerson, Michael O., and Christian Smith. *Divided by Faith: Evangelical Religion and the Problem of Race in America*. New York: Oxford University Press, 2001.

Emerson, Michael O., Christian Smith, and David Sikkink. "Equal in Christ, but Not in the World: White Conservative Protestants and Explanations of Black-White Inequality." *Social Problems* 46, no. 3 (1999): 398–417.

Epstein, Steven. *Inclusion: The Politics of Difference in Medical Research*. Chicago: University of Chicago Press, 2007.

Erlenmeyer-Kimling, Louise, and Lissy F. Jarvik. "Genetics and Intelligence: A Review." *Science* 142, no. 3598 (1963): 1477–79.

Fabian, Ann. *Skull Collectors: Race, Science, and America's Unburied Dead*. Chicago: University of Chicago Press, 2010.

Fagundes, Nelson J. R., Ricardo Kanitz, Roberta Eckert, Ana C. S. Valls, Mauricio R. Bogo, Francisco M. Salzano, David Glenn Smith, Wilson A Silva Jr., Marco A. Zago, Andrea K. Ribeiro-Dos-Santos, Sidney E. B. Santos, Maria Luiza Petzl-Erler, and Sandro L. Bonatto. "Mitochondrial Population Genomics Supports a Single Pre-Clovis Origin with a Coastal Route for the Peopling of the Americas." *American Journal of Human Genetics* 82, no. 3 (2008): 583–92.

Falconer, Douglas S. *Introduction to Quantitative Genetics*. New York: Ronald Press, 1960.

Farooq, Faheem, and John J. Strouse. "Disparities in Foundation and Federal Support and Development of New Therapeutics for Sickle Cell Disease and Cystic Fibrosis." *Blood* 132, supplement 1 (2018): 4687.

Fedyk, Mark. *The Social Turn in Moral Psychology*. Cambridge, MA: MIT Press, 2017.

Feldman, Marcus W., and Richard C. Lewontin. "Race, Ancestry, and Medicine." In *Revisiting Race in a Genomic Age*, ed. Barbara Koenig, Sandra Soo-Jin Lee, and Sarah S. Richardson, 89–101. New Brunswick, NJ: Rutgers University Press, 2008.

Félix, Suad El Burai, Cathy M. Bailey, and Hatice S. Zahran. "Asthma Prevalence among Hispanic Adults in Puerto Rico and Hispanic Adults of Puerto Rican Descent in the United States—Results from Two National Surveys." *Journal of Asthma* 52, no. 1 (2015): 3–9.

Ferber, Abby L. "The Construction of Race, Gender, and Class in White Supremacist Discourse." *Race, Gender and Class* 6, no. 3 (1999): 67–89.

Fisher, Ronald A. "The Causes of Human Variability." *Eugenics Review* 10, no. 4 (1919): 213–20.

———. "The Correlation between Relatives on the Supposition of Mendelian Inheritance." *Earth and Environmental Science Transactions of the Royal Society of Edinburgh* 52 (1919): 399–433.

———. "Eugenics, Academic and Practical." *Eugenics Review* 27 (1935): 95–100.

Ford, Chandra, and Nina T. Harawa. "A New Conceptualization of Ethnicity for Social Epidemiologic and Health Equity Research." *Social Science of Medicine* 71, no. 2 (2010): 251–58.

Foster-Hanson, E., and M. Rhodes. "Stereotypes as Prototypes in Children's Gender Concepts." *Developmental Science*, 26 (2023): e13345.

Foucault, Michel. "Questions of Method." In *Power: Essential Works of Foucault, 1954–1984*, ed. James D. Faubion, 223–38. New York: The New Press, [1978] 2000.

Fox, Dov. "Racial Classification in Assisted Reproduction." *Yale Law Journal* 118, no. 8 (2009): 1844–99.

Freedman, D. A. *Statistical Models and Causal Inference: A Dialogue with the Social Sciences.* New York: Cambridge University Press, 2010.

Freeman, David W. "Sperm Bank to Redheads: We Don't Want Your Semen." CBS News, September 10, 2011, https://cbsnews.com/news/sperm-bank-to-redheads-we-dont-want-your-semen.

Freese, Jeremy. "The Arrival of Social Science Genomics." *Contemporary Sociology* 47, no. 5 (2018): 524–36.

Freese, Jeremy, Ben Domingue, Sam Trejo, Kamil Sicinski, and Pamela Herd. "Problems with a Causal Interpretation of Polygenic Score Differences between Jewish and Non-Jewish Respondents in the Wisconsin Longitudinal Study." SocArXiv (2019). https://osf.io/preprints/socarxiv/eh9tq.

Friedman, Stanley. "Artificial Insemination with Donor Semen Mixed with Semen of the Infertile Husband." *Fertility and Sterility* 33, no. 2 (1980): 125–28.

Fuentes, Agustín. *Race, Monogamy, and Other Lies They Told You: Busting Myths about Human Nature.* 2nd ed. Oakland: University of California Press, 2022.

Fujimura, Joan H., Deborah A. Bolnick, Ramya Rajagopalan, Jay S. Kaufman, Richard C. Lewontin, Troy Duster, Pilar Ossorio, and Jonathan Marks. "Clines without Classes: How to Make Sense of Human Variation." *Sociological Theory* 32, no. 3 (2014): 208–27.

Fullwiley, Duana. "The Biologistical Construction of Race: 'Admixture' Technology and the New Genetic Medicine." *Social Studies of Science* 38, no. 5 (2008): 695–735.

Funk, C., and B. Kennedy. "The Politics of Climate." Pew Research Center, October 4, 2016. https://pewresearch.org/internet/wp-content/uploads/sites/9/2016/10/PS_2016.10.04_Politics-of-Climate_FINAL.pdf.

Gallagher, James. "'Huge Advance' in Fighting World's Biggest Killer." BBC, March 17, 2017. http://www.bbc.co.uk/news/health-39305640.

Gal'perin, P. Y. "On the Notion of Internalization." *Soviet Psychology* 5 (1967): 28–33.

Galton, Francis. *Essays in Eugenics.* London: Eugenics Education Society, 1909.

———. *Hereditary Genius.* London: Macmillan, 1869.

———. "Hereditary Talent and Character." *Macmillan's Magazine* 12, no. 68 and 70 (1865): 157–66 and 318–27.

Ganna, Andrea, Karin J. H. Verweij, Michel G. Nivard, Robert Maier, Robbee Wedow, Alexander S. Bush, Abdel Abdellaoui, Shengru Guo, J. Fah Sathirapongsasuti, 23andMe Research Team, et al. "Large-Scale GWAS Reveals Insights into the Genetic Architecture of Same-Sex Sexual Behavior." *Science* 365 (2019): 6456.

Garbus, Liz, dir. *Harry and Meghan.* Netflix, 2022. https://netflix.com/title/81439256.

Gasaway, Brantley W. *Progressive Evangelicals and the Pursuit of Social Justice.* Chapel Hill: University of North Carolina Press, 2014.

Geddes, Linda, and Michael Marshall. "Once, Twice, Thrice into the Americas." *New Scientist* 215, no. 2873 (2012): 12.

Gerbner, Katharine. *Christian Slavery: Conversion and Race in the Protestant Atlantic World.* Philadelphia: University of Pennsylvania Press, 2018.

Gibbon, Sarah, Ricardo Ventura Santos, and Mónica Sans, eds. *Racial Identities, Genetic Ancestry and Health in South America.* New York: Palgrave Macmillan, 2012.

Gibbons, Ann. "Genes Suggest Three Groups Peopled the New World." *Science* 337, no. 6091 (2012): 144.

———. "Geneticists Trace the DNA Trail of the First Americans." *Science* 259, no. 5093 (1993): 312–13.

Gilger, Jeffrey W. "Contributions and Promise of Human Behavioral Genetics." *Human Biology* 72, no. 1 (2000): 229–55.

Gillett, C. "The Metaphysics of Realization, Multiple Realizability, and the Special Sciences." *Journal of Philosophy* 100 (2023): 591–603.

Glasgow, Joshua, Sally Haslanger, Chike Jeffers, and Quayshawn Spencer. *What Is Race? Four Philosophical Views.* New York: Oxford University Press, 2019.

Globe Newswire. "Fertility Services Market Size Worth USD 46.06 Billion by 2030." February 17, 2023. https://globenewswire.com/en/news-release/2023/02/17/2610675/0/en/Fertility-Services-Market-Size-Worth-USD-46-06-Billion-by-2030-at-6-CAGR.

Goddard, Henry Herbert. *The Kallikak Family: A Study in the Heredity of Feeble-Mindedness.* New York: Macmillan, 1912.

Goetz, Rebecca Anne. *The Baptism of Early Virginia: How Christianity Created Race.* Baltimore: Johns Hopkins University Press, 2016.

Goldberg, Carey. "The Pandora's Box of Embryo Testing Is Officially Open." *Bloomberg,* May 26, 2022. https://www.bloomberg.com/news/features/2022-05-26/dna-testing-for-embryos-promises-to-predict-genetic-diseases.

Goodin, S. "Why Knowledge of the Internal Constitution Is Not the Same as Knowledge of the Real Essence and Why This Matters." *Southwest Philosophy Review* 14 (1998): 149–55.

Goodwin, Michele. "Reproducing Hierarchy in Commercial Intimacy." *Indiana Law Journal* 88 (2013): 1289–97.

Gordin, Michael. *A Well-Ordered Thing: Dmitrii Mendeleev and the Shadow of the Periodic Table.* Rev. ed. Princeton, NJ: Princeton University Press, 2018.

Gottfredson, Linda S. "Mainstream Science on Intelligence: An Editorial with 52 Signatories, History, and Bibliography." *Intelligence* 24, no. 1 (1994): 13–23.

Gould, Stephen J. *The Mismeasure of Man.* New York: W. W. Norton, 1981.

Gouveia, Mateus H., Amy R. Bentley, Thiago P. Leal, Eduardo Tarazona-Santos, Carlos D. Bustamante, Adebowale A. Adeyemo, Charles N. Rotimi, and Daniel Shriner.

"Unappreciated Subcontinental Admixture in Europeans and European Americans and Implications for Genetic Epidemiology Studies." *Nature Communications* 14 (2023): 6802.

Graham, Billy. "The Answers." Billy Graham Evangelistic Association, March 6, 2020. https://billygraham.org/answer/can-behavior-be-changed-by-altering-genes.

Graham, Franklin. "The Love That Conquers Racism." *Decision Magazine*, July 1, 2020. https://decisionmagazine.com/franklin-graham-the-love-that-conquers-racism.

Grant, Madison. *The Passing of the Great Race, or the Racial Basis of European History*. New York: Charles Scribner's Sons, 1916.

Gratton, Brian, and Emily Klancher Merchant. "*La Raza*: Mexicans in the United States Census." *Journal of Policy History* 28, no. 4 (2016): 537–67.

Graves, Joseph L., Jr. *The Emperor's New Clothes: Biological Theories of Race at the Millennium*. New Brunswick, NJ: Rutgers University Press, 2001.

———. "Favored Races in the Struggle for Life: Racism and the Speciation Concept." *Cold Spring Harbor Perspectives in Biology* 15, no. 8 (2023): 1–12.

———. "Out of Africa: Where Faith, Race, and Science Collide." In *Critical Approaches to Science and Religion*, ed. Myrna Perez Sheldon, Ahmed Ragab, and Terence Keel, 255–76. New York: Columbia University Press, 2023.

Graves, Joseph L., Jr., and Alan H. Goodman. *Racism, Not Race: Answers to Frequently Asked Questions*. New York: Columbia University Press, 2021.

Gravlee, C. C. "How Race Becomes Biology: Embodiment of Social Inequality." *American Journal of Physical Anthropology* 139, no. 1 (2009): 47–57.

Greeley, Henry T. "The Future of DTC Genomics and the Law." *Journal of Law, Medicine and Ethics* 48, no. 1 (2020): 151–60.

Greenberg, Joseph H., Christy G. Turner II, and Stephen L. Zegura. "The Settlement of the Americas: A Comparison of the Linguistic, Dental, and Genetic Evidence." *Current Anthropology* 27, no. 5 (1986): 477–97.

Greenspan, Ralph J. "The Origins of Behavioral Genetics." *Current Biology* 18, no. 5 (March 2008): 192–98.

Gregory, James N. *The Southern Diaspora: How the Great Migrations of Black and White Southerners Transformed America*. Chapel Hill: University of North Carolina Press, 2005.

Griesemer, James R. "A Data Journey through Dataset-Centric Population Genomics." In *Data Journeys in the Sciences*, ed. Sabina Leonelli and Niccolò Tempini, 146–67. Cham, Switzerland: Springer, 2020.

Griesemer, James R., and Carlos Andrés Barragán. "Re-situations of Scientific Knowledge: A Case Study of a Skirmish over Clusters vs. Clines in Human Population Genomics." *History and Philosophy of the Life Sciences* 44 (2022): article 16.

Grudem, Wayne, and John Piper. "The Danvers Statement: Recovering Biblical Manhood and Womanhood: A Response to Evangelical Feminism." Council on Biblical Manhood and Womanhood, 1997.

GWAS Diversity Monitor. https://gwasdiversitymonitor.com.

Haberstick, Brett C., Jeffrey M. Lessem, John K. Hewitt, Andrew Smolen, Christian J. Hopfer, Carolyn T. Halpern, Ley A. Killeya-Jones, Jason D. Boardman, Joyce Tabor, Ilene Siegler, Redford B. Williams, and Kathleen Mullan Harris. "MAOA Genotype,

Childhood Maltreatment, and Their Interaction in the Etiology of Adult Antisocial Behaviors." *Biological Psychiatry* 75, no. 1 (2014): 25–30.

Hacking, Ian. "Making Up People: Clinical Classifications." *London Review of Books*, 2006. https://www.lrb.co.uk/the-paper/v28/n16/ian-hacking/making-up-people.

Halder, Indrani, and Mark D. Shriver. "Measuring and Using Admixture to Study the Genetics of Complex Diseases." *Human Genomics* 1, no. 1 (2003): 52–62.

Ham, Ken, and Charles Ware. *One Race One Blood (Revised & Updated): The Biblical Answer to Racism*. Green Forest, AZ: New Leaf Publishing, 2019.

Hamer, Dean H., S. Hu, V. L. Magnuson, N. Hu, and A. M. Pattatucci. "A Linkage between DNA Markers on the X Chromosome and Male Sexual Orientation." *Science* 261, no. 5119 (1993): 321–27.

Hamilton, Michael S. "A Strange Love? Or: How White Evangelicals Learned to Stop Worrying and Love the Donald." In *Evangelicals: Who They Have Been, Are Now, and Could Be*, ed. Mark A. Noll, David W. Bebbington, and George M. Marsden, 217–27. Grand Rapids, MI: Eerdmans, 2019.

Harden, Kathryn Paige. *The Genetic Lottery: Why DNA Matters for Social Equality*. Princeton, NJ: Princeton University Press, 2021.

Harden, Kathryn Paige, Benjamin W. Domingue, Daniel W. Belsky, Jason D. Boardman, Robert Crosnoe, Margherita Malanchini, Michel Nivard, Elliot M. Tucker-Drob, and Kathleen Mullan-Harris. "Genetic Associations with Mathematics Tracking and Persistence in Secondary School." *Science of Learning* 5, no. 1 (2020). https://www.nature.com/articles/s41539-020-0060-2.

Harris, Cheryl I. "Whiteness as Property." *Harvard Law Review* 106, no. 8 (1993): 1707–91.

Harrison, Laura. "The Woman or the Egg? Race in Egg Donation and Surrogacy Databases." *Genders* 58 (Fall 2013): 24.

Harvard University. "Race in a Genetic World." *Harvard Magazine*, May–June 2008. https://harvardmazagine.com/2008/05/race-in-a-genetic-world-html.

Hattam, Victoria. "Ethnicity and the Boundaries of Race: Rereading Directive 15." *Daedalus* 134, no. 1 (2005): 61–69.

Hatter, Jonathan J. "Slavery and the Enslaved in the Roman World, the Jewish World, and the Synoptic Gospels." *Currents in Biblical Research* 20, no. 1 (2021): 97–127.

Hauser, Robert M. "Educational Stratification in the United States." *Sociological Inquiry* 40, no. 2 (1970): 102–29.

Havstad, Joyce C. "Sensational Science, Archaic Hominin Genetics, and Amplified Inductive Risk." *Canadian Journal of Philosophy* 5, no. 3 (2022): 295–320.

Hawkins, J. Russell. *The Bible Told Them So: How Southern Evangelicals Fought to Preserve White Supremacy*. New York: Oxford University Press, 2021.

Haynes, Stephen R. *Noah's Curse: The Biblical Justification of American Slavery*. New York: Oxford University Press, 2002.

Henn, Brenna, Emily Klancher Merchant, Anne O'Connor, and Tina Rulli. "Why DNA Is No Key to Social Equality: On Kathryn Paige Harden's *The Genetic Lottery*." *LA Review of Books*, September 21, 2021. https://lareviewofbooks.org/article/why-dna-is-no-key-to-social-equality-on-kathryn-paige-hardens-the-genetic-lottery.

Herrnstein, Richard, and Charles Murray. *The Bell Curve: Intelligence and Class Structure in American Life*. New York: Free Press, 1994.

Hickman, Christine B. "The Devil and the One Drop Rule: Racial Categories, African Americans, and the U.S. Census." *Michigan Law Review* 95, no. 5 (March 1997): 1161–265.

Higginbotham, A. Leon, and Barbara A. Kopytoff. "Racial Purity and Interracial Sex in the Law of Colonial and Antebellum Virginia." *Georgetown Law Journal* 77 (1988): 1967–2029.

Hill, W. David, Neil M. Davies, Stuart J. Ritchie, Nathan G. Skene, Julien Bryois, Steven Bell, Emanuele Di Angelantonio, David J. Roberts, Shen Xueyi, Gail Davies, David C. M. Liewald, David J. Porteous, Caroline Hayward, Adam S. Butterworth, Andrew M. McIntosh, Catharine R. Gale, and Ian J. Deary. "Genome-Wide Analysis Identifies Molecular Systems and 149 Genetic Loci Associated with Income." *Nature Communications* 10 (2019): 5741.

Hochschild, Jennifer, and Brenna Marea Powell. "Racial Reorganization and the United States Census 1850–1930: Mulattoes, Half-Breeds, Mixed Parentage, Hindoos, and the Mexican Race." *Studies in American Political Development* 22, no. 1 (2008): 59–96.

Hoffman, Kelly M., Sophie Trawalter, Jordan R. Axt, and M. Normal Oliver. "Racial Bias in Pain Assessment and Treatment Recommendations, and False Beliefs about Biological Differences between Blacks and Whites." *Proceedings of the National Academy of Sciences* 113, no. 16 (2016): 4296–301.

Holden, Michael. "Harry and Meghan Decry 'Pain and Suffering' of Women Brought into UK Royal Family." Reuters, December 8, 2022. https://www.reuters.com/lifestyle/british-royals-brace-harry-meghans-netflix-broadside-2022-12-08/.

Holtz, Robert Lee. "Early Americans Arrived in Three Waves." *Wall Street Journal*, July 12, 2012.

Holzinger, Karl J. "The Relative Effect of Nature and Nurture on Twin Differences." *Journal of Educational Psychology* 20, no. 4 (1929): 241–48.

HoSang, Daniel Martinez. "On Racial Speculation and Racial Science: A Response to Shiao et al." *Sociological Theory* 32, no. 3 (2014): 228–43.

Howlett, Peter, and Mary S. Morgan, eds. *How Well Do Facts Travel? The Dissemination of Reliable Knowledge*. Cambridge: Cambridge University Press, 2011.

Hudson, Kathy, Gail Javitt, Wylie Burke, and Peter Byers. "ASHG Social Issues Committee. ASHG Statement on Direct-to-Consumer Genetic Testing in the United States." *American Journal of Human Genetics* 81, no. 3 (2007): 635–75.

Hume, Brad D. "Quantifying Characters: Polygenist Anthropologists and the Hardening of Heredity." *Journal of the History of Biology* 41, no. 1 (2007): 119–58.

Hummel, Susanne. *Ancient DNA Typing: Methods, Strategies and Applications*. Berlin: Springer-Verlag, 2003.

Hunt, E. A., and W. H. Barton. "The Inconstancy of Physique in Adolescent Boys and Other Limitations of Somatotyping." *American Journal of Physical Anthropology* 17 (1959); 27–35.

Hyde, Marina. "Meghan Markle Is the Perfect Fit for Our New Touchy-Feely Royal Family." *Guardian*, September 7, 2017. https://www.theguardian.com/lifeandstyle/lostinshowbiz/2017/sep/07/meghan-markle-perfect-fit-touchy-feely-royal-family-prince-harry.

Ifekwunigwe, Jayne O., ed. *'Mixed Race' Studies: A Reader*. New York: Routledge, 2004.

Ikemoto, Lisa C. "Assisted Reproductive Technology Use among Neighbours: Commercialization Concerns in Canada and the United States, in the Global Context." In *Regulating Creation: The Law, Ethics, and Policy of Assisted Human Reproduction*, ed. Trudo Lemmens,

Andrew Flavell Martin, Cheryl Milne, and Ian B. Lee, 253–73. Toronto: Toronto University Press, 2017.

———. "Infertile by Force and Federal Complicity: The Story of *Relf v. Weinberger*." In *Women and the Law Stories*, ed. Elizabeth M. Schneider and Stephanie Wildman, chap. 5. New York: Foundation Press / Thomson Reuters, 2011.

———. "Racial Disparities in Health Care and Cultural Competency." *Saint Louis University Law Journal* 48, no. 1 (2003): 92–93.

———. "Reproductive Tourism." In *Beyond Bioethics: Toward a New Biopolitics*, ed. Osagie K. Obasogie and Marcy Darnovsky, 339–49. Oakland: University of California Press, 2018.

Ireland, Alleyne. *Democracy and the Human Equation*. New York: E. P. Dutton, 1921.

James, Michael. "Race." *Stanford Encyclopedia of Philosophy*. 2020. https://plato.stanford.edu/entries/race.

Jeanson, Nathaniel. *Traced: Human DNA's Big Surprise*. Green Forest, AZ: New Leaf Publishing Group, 2022.

Jencks, Christopher, Marshall Smith, Henry Acland, Mary Jo Bane, David Cohen, Herbert Gintis, Barbara Heyns, and Stephan Michelson. *Inequality: A Reassessment of the Effect of Family and Schooling in America*. New York: Basic Books, 1972.

Jensen, Arthur R. "Estimation of the Limits of Heritability of Traits by Comparison of Monozygotic and Dizygotic Twins." *Proceedings of the National Academy of Sciences* 58, no. 1 (1967): 149–56.

———. "How Much Can We Boost IQ and Scholastic Achievement?" *Harvard Educational Review* 39, no. 1 (1969): 1–124.

Jobling, Mark, Edward Hollox, Toomas Kivisild, and Chris Tyler-Smith. *Human Evolutionary Genetics*. 2nd ed. New York: Garland Science, [2004] 2014.

Johfre, Sasha, Aliya Saperstein, and Jill A. Hollenbach. "Measuring Race and Ancestry in the Age of Genetic Testing." *Demography* 58, no. 3 (2021): 785–810.

Johnston, Jeff. "Male and Female He Created Them: Genesis and God's Design of Two Sexes." *Focus on the Family*, September 13, 2015.

Jones, Kevin. "Report on Slavery and Racism in the History of the Southern Baptist Theological Seminary," December 2018. https://digitalcommons.cedarville.edu/education_publications/108.

Jones, Robert P. *White Too Long: The Legacy of White Supremacy in American Christianity*. New York: Simon and Schuster, 2021.

Jordan, Winthrop D. "Historical Origins of the One-Drop Racial Rule in the United States." *Journal of Critical Mixed Race Studies* 1, no. 1 (2014): 98–132.

Journal of the American Medical Association. "Instructions for Authors: Reporting Demographic Information for Study Participants." https://jamanetwork.com/journals/jama/pages/instructions-for-authors#SecReportingRace/Ethnicity.

Juerst, J., V. Shibaev, and E. O. W. Kirkegaard. "A Genetic Hypothesis for American Race/Ethnic Differences in Mean g: A Reply to Warne (2021) with Fifteen New Empirical Tests Using the ABCD Dataset." *Mankind Quarterly* 63, no. 4 (2023): 527–600.

Junior, Nyasha. "The Mark of Cain and White Violence." *Journal of Biblical Literature* 139, no. 4 (2020): 661–73.

Kaback, Michael, Joyce Lim-Steele, Deepti Dabholkar, David Brown, Nancy Levy, and Karen Zeiger. "Tay-Sachs Disease—Carrier Screening, Prenatal Diagnosis, and the Molecular Era: An International Perspective, 1970 to 1993." *JAMA* 270, no. 19 (1993): 2307–15.

Kahn, Jonathan. 2014. "How Not to Talk about Race and Genetics." *BuzzFeed News*, March 30, 2018. https://buzzfeednews.com/article/bfopinion/race-genetics-david-reich.

———. *Race in a Bottle: The Story of BiDil and Racialized Medicine in a Post-Genomic Age.* New York: Columbia University Press, 2014.

———. "'When Are You From?' Time, Space, and Capital in the Molecular Reinscription of Race." *British Journal of Sociology* 66, no. 1 (2015): 68–75.

Kamin, Leon. *The Science and Politics of IQ.* New York: Halsted Press, 1974.

Kaufman, Jay S, Lena Dolman, Dinela Rushani, and Richard S. Cooper. "The Contribution of Genomic Research to Explaining Racial Disparities in Cardiovascular Disease: A Systematic Review." *American Journal of Epidemiology* 181, no. 7 (2015): 464–72.

Keaney, Jaya. "The Racializing Womb: Surrogacy and Epigenetic Kinship." *Science, Technology, and Human Values* 47, no. 6 (2019): 1157–79.

Keel, Terence. *Divine Variations: How Christian Thought Became Racial Science.* Stanford, CA: Stanford University Press, 2018.

———. "Religion, Polygenism and the Early Science of Human Origins." *History of the Human Sciences* 26, no. 2 (April 2013): 3–32.

Keller, Timothy. "The Bible and Race." *Life in the Gospel*, Spring 2020. https://quarterly .gospelinlife.com/the-bible-and-race.

———. "The Sin of Racism." *Life in the Gospel*, Summer 2020. https://quarterly.gospelinlife .com/the-sin-of-racism.

Kevles, Daniel J. *In the Name of Eugenics: Genetics and the Uses of Human Heredity.* Cambridge, MA: Harvard University Press, 1995.

Kidd, Colin. *The Forging of Races: Race and Scripture in the Protestant Atlantic World, 1600–2000.* Cambridge: Cambridge University Press, 2006.

King, Miriam, and Steven Ruggles. "American Immigration, Fertility, and Race Suicide at the Turn of the Century." *Journal of Interdisciplinary History* 20, no. 3 (1990): 347–69.

Kirk, R. "Nonreductive Physicalism and Strict Implication." *Australasian Journal of Philosophy* 79 (2001): 544–52.

Kleeman, Jenny. "America's Premier Pronatalists on Having 'Tons of Kids' to Save the World: 'There Are Going to Be Countries of Old People Starving to Death.'" *Guardian*, May 25, 2024. https://www.theguardian.com/lifeandstyle/article/2024/may/25/american-pronatalists -malcolm-and-simone-collins.

Klein, Richard G. *The Human Career: Human Biological and Cultural Origins.* Chicago: University of Chicago Press, [1989] 2009.

Klineberg, Otto. *Negro Intelligence and Selective Migration.* New York: Columbia University Press, 1935.

———. *Race Differences.* New York: Harper and Brothers, 1935.

Kluchin, Rebecca M. *Fit to Be Tied: Sterilization and Reproductive Rights in America, 1950–1980.* New Brunswick, NJ: Rutgers University Press, 2011.

Koenig, Barbara A., Sandra Soo-Jin Lee, and Sarah S. Richardson, eds. *Revisiting Race in a Genomic Age.* New Brunswick, NJ: Rutgers University Press, 2008.

Kopelson, Heather Miyano. *Faithful Bodies: Performing Religion and Race in the Puritan Atlantic*. New York: NYU Press, 2014.

Korlach, Jonas. "We Need More Diversity in Our Genomic Databases." *Scientific American: Voices*, December 4, 2018. https://blogs.scientificamerican.com/voices/we-need-more-diversity-in-our-genomic-databases.

Korsh, Aaron, creator. *Suits*. Netflix, 2011–19. https://imdb.com/title/tt1632701.

Korunes, Katherine L., and Amy Goldberg. "Human Genetic Admixture." *PLOS Genetics* 17, no. 3 (2021): e1009347.

Kramer, Karen L. "The Human Family—Its Evolutionary Context and Diversity." *Social Sciences* 10, no. 6 (2021). https://doi.org/10.3390/socsci10060191.

Krishnamurthy, Anna Purna. "In 1968, These Activists Coined the Term 'Asian-American'—and Helped Shape Decades of Advocacy." *Time*, May 22, 2020. https://time.com/5837805/aisan-american-history.

Ladyman, J. "What Is Structural Realism?" *Studies in History and Philosophy of Science* 29 (1998): 409–24.

Lange, Lynda. "Rousseau and Modern Feminism." *Social Theory and Practice* 7 (1981): 245–77.

Larson, Edward J. *Sex, Race, and Science: Eugenics in the Deep South*. Baltimore: Johns Hopkins University Press, 1996.

Lasker, Jordan, Bryan J. Pesta, John G. R. Fuerst, and Emil O. W. Kirkegaard. "Global Ancestry and Cognitive Ability." *Psych* 1, no. 1 (2019): 431–59.

Latour, Bruno. *On the Modern Cult of the Factish Gods*. Durham, NC: Duke University Press, 2010.

Laughlin, Harry Hamilton. *Eugenical Sterilization in the United States*. Chicago: Psychopathic Laboratory of the Municipal Court of Chicago, 1922.

Lawton, Georgina. "I Usually Disparage the Royals, but Meghan Markle Has Changed That." *Guardian*, October 21, 2017. https://www.theguardian.com/lifeandstyle/2017/oct/21/royals-meghan-markle-uk-change-race-relations-prince-harry.

Lee, James J. "The Heritability and Persistence of Social Class in England." *Proceedings of the National Academy of Sciences* 120, no. 29 (2023): e2309250120.

Lee, James J., Robbee Wedow, Aysu Okbay, Edward Kong, Omeed Maghzian, Meghan Zacher, Tuan Anh Nguyen-Viet, Peter Bowers, Julia Sidorenko, Richard Karlsson Linnér, et al. "Gene Discovery and Polygenic Prediction from a 1.1-Million-Person-GWAS of Educational Attainment." *Nature Genetics* 50, no. 8 (August 2018): 1112–21.

Lee, Sandra Soo-Jin, Deborah A. Bolnick, Troy Duster, Pilar Ossorio, and Kimberly TallBear. "The Illusive Gold Standard in Genetic Ancestry Testing." *Science* 325, no. 5936 (2009): 38–39.

Lee, Sharon M. "Racial Classifications in the U.S. Census: 1890–1990." *Ethnic and Racial Studies* 16, no. 1 (1993): 75–94.

Lemos, Tracy M. "Physical Violence and the Boundaries of Personhood in the Hebrew Bible." *Hebrew Bible and Ancient Israel (HeBAI)* 2 (2013): 500–531.

Leonard, Thomas C. "Eugenics and Economics in the Progressive Era." *Journal of Economic Perspectives* 19, no. 4 (2005): 207–24.

Leonelli, Sabina, and Niccoló Tempini, eds. *Data Journeys in the Sciences*. Cham, Switzerland: Springer, 2020.

Lewis, Anna C. F., Santiago J. Molina, Paul S. Appelbaum, Bege Dauda, Anna Di Rienzo, Agustín Fuentes, Stephanie M. Fullerton, Nanibaa' A. Garrison, Nayanika Ghosh, et al. "Getting Genetic Ancestry Right for Science and Society." *Science* 376, no. 6590 (2022): 250–52.

Lewontin, Richard C. "Annotation: The Analysis of Variance and the Analysis of Causes." *American Journal of Human Genetics* 26 (1974): 400–411.

———. "The Apportionment of Human Diversity." In *Evolutionary Biology*, ed. Theodosius Dobzhansky, Max K. Hecht, and William C. Steere, vol. 6, 381–98. New York: Springer, 1972.

———. *Biology as Ideology: The Doctrine of DNA*. New York: Harper Perennial, [1991] 1992.

———. "Race and Intelligence." *Bulletin of the Atomic Scientists* 26, no. 3 (1970): 2–8.

Lifeline Children's Services. "Stepping into a Multi-Cultural World: Racial Reconciliation and the Gospel." https://lifelinechild.org/stepping-into-a-multi-cultural-world-racial -reconciliation-the-gospel.

Lim, Regine M., Ari J. Silver, Maxwell J. Silver, Carlos Borroto, Brett Spurrier, Tanya C. Petrossian, Jessica L. Larson, and Lee M. Silver. "Targeted Mutation Screening Panels Expose Systematic Population Bias in Detection of Cystic Fibrosis Risk." *Genetics in Medicine* 18, no. 2 (2016): 174–79.

Lindqvist, Charlotte, and Om P. Rajora. *Paleogenomics: Genome-Scale Analysis of Ancient DNA*. Cham, Switzerland: Springer, 2019.

Lipka, Michael, and Patricia Tevington. "Attitudes about Transgender Issues Vary Widely Among Christians, Religious 'Nones' in U.S." Pew Research Center, July 7, 2022. https:// pewresearch.org/short-reads/2022/07/07/attitudes-about-transgender-issues-vary- widely-among-christians-religious-nones-in-u-s.

Lippmann, Walter. "The Abuse of the Tests." *New Republic* 32, no. 415 (1922): 297–98.

———. "A Future for the Tests." *New Republic* 32, no. 417 (1922): 9–11.

———. "The Mental Age of Americans." *New Republic* 32, no. 412 (1922): 213–15.

———. "The Mystery of the 'A' Men." *New Republic* 32, no. 413 (1922): 246–48.

———. "The Reliability of Intelligence Tests." *New Republic* 32, no. 414 (1922): 275–77.

———. "Tests of Hereditary Intelligence." *New Republic* 32, no. 416 (1922): 328–30.

Lipsitz, George. "The Possessive Investment in Whiteness: Racialized Social Democracy and the 'White' Problem in American Studies." *American Quarterly* 47, no. 3 (September 1995): 369–87.

Locke, John. *An Essay Concerning Human Understanding*. Indianapolis: Hackett Publishing Company, 1996.

Lofton, Kathryn. "Religious History as Religious Studies." *Religion* 42, no. 3 (2012): 383–94.

Lombardo, Paul. *Three Generations, No Imbeciles: Eugenics, the Supreme Court and Buck v. Bell*. Baltimore: Johns Hopkins University Press, 2008.

Lombrozo, T. "Causal-Explanatory Pluralism: How Intentions, Functions, and Mechanisms Influence Causal Ascriptions." *Cognitive Psychology* 61 (2010): 303–32.

Lowe, E. J. "Locke on Real Essence and Water as a Natural Kind: A Qualified Defense." *Aristotelian Society Supplementary Volume* 85 (2011): 1–19.

Löwy, Ilana. "How Genetics Came to the Unborn: 1960–2000." *Studies in History and Philosophy of Science Part C: Studies in History and Philosophy of Biological and Biomedical Sciences* 47 (2014): 154–62.

Ludmerer, Kenneth M. "Genetics, Eugenics, and the Immigration Restriction Act of 1924." *Bulletin of the History of Medicine* 46, no. 1 (1972): 59–81.

Lum, Kathryn Gin. *Heathen: Religion and Race in American History*. Cambridge, MA: Harvard University Press, 2022.

Lupton, Deborah. *The Quantified Self: A Sociology of Self-Tracking*. Cambridge: Polity Press, 2016.

Luse, Christopher A. "Slavery's Champions Stood at Odds: Polygenesis and the Defense of Slavery." *Civil War History* 53, no. 4 (2007): 379–412.

Lush, Jay Laurence. *Animal Breeding Plans*. Ames, IA: Collegiate Press, 1943.

MacArthur, John. "Act like Men." Grace to You, June 21, 2020. https://gty.org/library/sermons-library/81-82/act-like-men.

———. "The Sin of Noah." Grace to You, June 17, 2001. https://gty.org/library/sermons-library/90-264/the-sin-of-noah.

Maghbouleh, Neda, Ariela Schacter, and René D. Flores. "Middle Eastern and North African Americans May Not Be Perceived, nor Perceive Themselves, to Be White." *Proceedings of the National Academy of Sciences* 119, no. 7 (2020): e2117940119.

Mahdawi, Arwa. "'Hipster Eugenics': Why Is the Media Cosying Up to People Who Want to Build a Superior Race?" *Guardian*, April 21, 2023. https://theguardian.com/lifeandstyle/2023/apr/20/pro-natalism-babies-global-population-genetics.

Maltby, Clive, dir. *Journey of Man*. 2003; Alexandria: PBS Home Video and National Geographic Society.

Mamo, Laura. *Queering Reproduction: Achieving Pregnancy in the Age of Technoscience*. Durham, NC: Duke University Press, 2007.

Manrai, Arjun K., et al. "Genetic Misdiagnoses and the Potential for Health Disparities." *New England Journal of Medicine* 375 (2016): 655–65.

Marks, Jonathan. "Race: Past, Present, and Future." In *Revisiting Race in a Genomic Age*, ed. Barbara A. Koenig, Sandra Soo-Jin Lee, and Sarah S. Richardson, 20–38. New Brunswick, NJ: Rutgers University Press, 2008.

Martin, Alicia R., Christopher R. Gignoux, Raymond K. Walters, Genevieve L. Wojcik, Benjamin M. Neale, Simon Gravel, Mark J. Daly, Carlos D. Bustamante, and Eimear E. Kenney. "Human Demographic History Impacts Genetic Risk Prediction across Diverse Populations." *American Journal of Human Genetics* 100, no. 40 (2017): 635–49.

Martin, Alicia R., Masahiro Kanai, Yoichiro Kamatani, Yukinori Okada, Benjamin M. Neale, and Mark J. Daly. "Clinical Use of Current Polygenic Risk Scores May Exacerbate Health Disparities." *Nature Genetics* 51, no. 4 (2019): 584–91.

Martineau, Paris. "Inside the Quietly Lucrative Business of Donating Human Eggs." *Wired*, April 23, 2019. https://www.wired.com/story/inside-lucrative-business-donating-human-eggs.

Mathieson, Iain, and Aylwyn Scally. "What Is Ancestry?" *PLoS Genetics* 16, no. 3 (2020): e1008624.

Maxmen, Amy. "Controversial 'Gay Gene' App Provokes Fears of a Genetic Wild West." *Nature* 574 (2019): 609–10.

Maxwell, Angie, Pearl Forde Dowe, and Todd Shields. "The Next Link in the Chain Reaction: Symbolic Racism and Obama's Religious Affiliation." *Social Science Quarterly* 94, no. 2 (2013): 321–43.

Mays, Vickie M., Ninez A. Ponce, Donna L. Washington, and Susan D. Cochran. "Classification of Race and Ethnicity: Implications for Public Health." *Annual Review of Public Health* 24, no. 1 (2003): 83–110.

McCloud, Sean. *Divine Hierarchies: Class in American Religion and Religious Studies.* Chapel Hill: University of North Carolina Press, 2009.

McDaniel, Charles. "John A. Ryan and the American Eugenics Society: A Model for Christian Engagement in the Age of Consumer Eugenics." *Journal of Religion and Society* 22 (2020): 1–22.

McDougall, William. *Is America Safe for Democracy?* New York: Charles Scribner's Sons, 1921.

McKeigue, Paul M. "Mapping Genes That Underlie Ethnic Differences in Disease Risk: Methods for Detecting Linkage in Admixed Populations, by Conditioning on Parental Admixture." *American Journal of Human Genetics* 63, no. 1 (1998): 241–51.

———. "Prospects for Admixture Mapping of Complex Traits." *American Journal of Human Genetics* 76, no. 1 (2005): 1–7.

Meehl, P. E. "Theoretical Risks and Tabular Asterisks: Sir Karl, Sir Ronald, and the Slow Progress of Soft Psychology." In *The Restoration of Dialogue: Readings in the Philosophy of Clinical Psychology*, vol. 654, ed. R. B. Miller, 523–55. American Psychological Association, 1992.

———. "Theory-Testing in Psychology and Physics: A Methodological Paradox." *Philosophy of Science* 34 (1967): 103–15.

Mehta, Sharan Kaur, Rachel C. Schneider, and Elaine Howard Ecklund. "'God Sees No Color' So Why Should I? How White Christians Produce Divinized Colorblindness." *Sociological Inquiry* 92, no. 2 (2022): 623–46.

Meltzer, David J. *First Peoples in a New World: Colonizing Ice Age America.* Berkeley: University of California Press, 2009.

Merchant, Emily Klancher. *Building the Population Bomb.* New York: Oxford University Press, 2021.

———. "Environmental Malthusianism and Demography." *Social Studies of Science* 52, no. 4 (2022): 536–60.

———. "Hold Science to Higher Standards on Racism." *STAT*, June 30, 2022. https://statnews.com/2022/06/20/hold-science-to-higher-standards-on-racism.

———. "Of DNA and Demography." In *Recent Trends in Demographic Data*, ed. Parfait M. Eloundou-Enyegue. London: IntechOpen, 2023. https://intechopen.com/online-first/1129796.

Meyer, Michelle N., Paul S. Appelbaum, Daniel J. Benjamin, Shawneequa L. Callier, Nathaniel Comfort, Dalton Conley, Jeremy Freese, Nanibaa' A. Garrison, Evelynn M. Hammonds, K. Paige Harden, Sandra Soo-Jin Lee, Alicia R. Martin, Daphne Oluwaseun Martschenko, Benjamin M. Neale, Rohan H. C. Palmer, James Tabery, Eric Turkheimer, Patrick Turley, and Erik Parens. "Wrestling with Social and Behavioral Genomics: Risks, Potential Benefits, and Ethical Responsibility." *Hastings Center Report* 53, no. S1 (2023): S2–49.

Meyer, Michelle N., Tammy Tan, Daniel J. Benjamin, David Laibson, and Patrick Turley. "Public Views on Polygenic Screening of Embryos." *Science* 379, no. 6632 (2023): 541–43.

Millstein, Roberta. "Thinking about Populations and Races in Time." *Studies in History and Philosophy of Biological and Biomedical Sciences* 52 (2015): 5–11.

Mitchell, Paul Wolff. "The Fault in His Seeds: Lost Notes to the Case of Bias in Samuel George Morton's Cranial Race Science." *PLOS Biology* 16, no. 10 (October 4, 2018): e2007008.

Mohamed, Nadifa. "As Meghan Has Learned, the Monarchy Is Still Built on Breeding, Ancestry and Caste." *Guardian*, March 8, 2021. https://www.theguardian.com/commentisfree /2021/mar/08/meghan-monarchy-breeding-ancestry-caste.

Mohler, Albert. "The Briefing." August 30, 2019. https://albertmohler.com/2019/08/30 /briefing-8-30-19.

———. "The Briefing." June 24, 2020. https://albertmohler.com/2020/06/24/briefing -6-24-20.

———. "The Briefing." June 17, 2022. https://albertmohler.com/2022/06/17/briefing-6-17-22.

———. "The Heresy of Racial Superiority—Confronting the Past, and Confronting the Truth," June 23, 2015. https://albertmohler.com/2015/06/23/the-heresy-of-racial-superiority -confronting-the-past-and-confronting-the-truth.

———. "Is Homosexuality in the Genes?" April 4, 2004. https://albertmohler.com/2004 /04/15/is-homosexuality-in-the-genes.

Moll, Tessa. "Making a Match: Curating Race in South African Gamete Donation." *Medical Anthropology* 38, no. 7 (2019): 588–602.

Montoya, Michael. *Making the Mexican Diabetic: Race, Science, and the Genetics of Inequality*. Oakland: University of California Press, 2011.

Morning, Ann. "Does Genomics Challenge the Social Construction of Race?" *Sociological Theory* 32, no. 3 (2014): 189–207.

Mostafavi, Makhamanesh, Arbel Harpak, Ipsita Agarwal, Dalton Conley, Jonathan K. Pritchard, and Molly Przeworski. "Variable Prediction Accuracy of Polygenic Scores within an Ancestry Group." *eLife* 9 (2020): e48376.

Mullin, Joe. "White Woman Sues Sperm Bank—Again—After Getting Black Man's Sperm." *Ars Technica*, April 25, 2016. https://arstechnica.com/tech-policy/2016/04/white-woman -sues-sperm-bankagainafter-getting-black-mans-sperm.

Nash, Catherine. *Genetic Geographies: The Trouble with Ancestry*. Minneapolis: University of Minnesota Press, 2015.

National Academies of Sciences, Engineering and Medicine. *Using Population Descriptors in Genetics and Genomics Research: A New Framework for an Evolving Field*. Washington, DC: National Academies Press, 2023. https://nap.nationalacademies.org/catalog/26902 /using-population-descriptors-in-genetics-and-genomics-research-a-new.

Nature. "Editorial: Why *Nature* Is Updating Its Advice to Authors on Reporting Race or Ethnicity." *Nature* 616, no. 7956 (2023): 219.

Nelkin, Dorothy, and M. Susan Lindee. *The DNA Mystique: The Gene as a Cultural Icon*. Ann Arbor: University of Michigan Press, 1995.

New York Times. "Race, Genetics and a Controversy." April 2, 2018. https://nytimes.com /2018/04/02/opinion/genes-race.html.

———. "Text of the White House Statements on the Human Genome Project." June 27, 2000. https://archive.nytimes.com/www.nytimes.com/library/national/science/062700sci -genome-text.html.

Newman, Alyssa. "Mixing and Matching: Sperm Donor Selection for Interracial Lesbian Couples." *Medical Anthropology* 38, no. 8 (2019): 710–24.

Nobles, Melissa. *Shades of Citizenship: Race and the Census in Modern Politics.* Stanford, CA: Stanford University Press, 2000.

Noll, Mark A. *God and Race in American Politics: A Short History.* Princeton, NJ: Princeton University Press, 2008.

Nozick, Robert. *Anarchy, State, and Utopia.* New York: Basic Books, 1974.

O'Brien, Stephen J. "Stewardship of Human Biospecimens, DNA, Genotype, and Clinical Data in the GWAS Era." *Annual Review of Genomics and Human Genetics* 10 (2009): 193–209.

Office of Management and Budget. "Statistical Directive No. 15: Race and Ethnic Standards for Federal Statistics and Administrative Reporting." https://wonder.cdc.gov/wonder /help/populations/bridged-race/directive15.html.

Ogletree, Charles J., Jr. "The Significance of *Brown*." *Judicature* 88 (2004): 66–72.

Okbay, Aysu, Jonathan P. Beauchamp, Mark Alan Fontana, James J. Lee, Tune H. Pers, Cornelius A. Reitveld, Patrick Turley, Guo-Bo Chen, Valur Emilsson, S. Fleur W. Meddens, et al. "Genome-Wide Association Study Identifies 74 Loci Associated with Educational Attainment." *Nature* 533 (2016): 539–42.

Okbay, Aysu, Yeda Wu, Nancy Wang, Hariharan Jayashankar, Michael Bennett, Seyed Moeen Nehzati, Julia Sidorenko, Hyeokmoon Kweon, Grant Goldman, Tamera Gjorgjieva, et al. "Polygenic Prediction of Educational Attainment within and between Families from Genome-Wide Association Analyses in 3 Million Individuals." *Nature Genetics* 54 (2022): 437–49.

Omi, Michael, and Howard Winant. *Racial Formation in the United States.* New York: Routledge, 2014.

Onishi, Norimitsu. "Switched at Birth, Two Canadians Discover Their Roots at 67." *New York Times*, August 2, 2023. https://nytimes.com/2023/08/02/world/canada/canada-men -switched-at-birth.html.

Onishi, Y., and D. Serpico. "Homeostatic Property Cluster Theory without Homeostatic Mechanisms: Two Recent Attempts and Their Costs." *Journal for General Philosophy of Science* 53 (2022): 61–82.

Oppenheim, Paul, and Hilary Putnam. "Unity of Science as a Working Hypothesis." In *Concepts, Theories, and the Mind-Body Problem: Minnesota Studies in the Philosophy of Science*, ed. Herbert Feigl, Michael Criven, and Grover Maxwell, 3–36. Minneapolis: University of Minnesota Press, 1958.

Orcés, Diana. "Black, White, and Born Again: How Race Affects Opinions among Evangelicals." Public Religion Research Institute, February 17, 2021. https://prri.org/spotlight /black-white-and-born-again-how-race-affects-opinions-among-evangelicals.

Ordover, Nancy. *American Eugenics: Race, Queer Anatomy, and the Science of Nationalism.* Minneapolis: University of Minnesota Press, 2003.

Osborn, Frederick Henry. "Development of a Eugenic Philosophy." *American Sociological Review* 2, no. 3 (1937): 389–97.

Osborne, Richard H., and Barbara T. Osborne. "The Founding of the Behavior Genetics Association, 1966–1971." *Social Biology* 46, no. 3–4 (1999): 207–18.

Outram, Simon M., and George T. H. Ellison. "Anthropological Insights into the Use of Race/Ethnicity to Explore Genetic Determinants of Disparities in Health." *Journal of Biosocial Science* 38, no. 1 (2005): 83–102.

———. "Improving the Use of Race/Ethnicity in Genetic Research: A Survey of Instructions to Authors in Genetics Journals." *Science Education* 29, no. 3 (2006): 78–81.

Pande, Amrita. "Mix or Match? Transnational Fertility Industry and White Desirability." *Medical Anthropology* 40, no. 4 (2021): 335–47.

Panofsky, Aaron. "From Behavior Genetics to Postgenomics." In *Postgenomics: Perspectives on Biology after the Genome*, ed. Sarah Richardson and Hallam Stevens, 150–73. Durham, NC: Duke University Press, 2015.

———. *Misbehaving Science: Controversy and the Development of Behavior Genetics.* Chicago: University of Chicago Press, 2014.

Panofsky, Aaron, Kushan Dasgupta, and Nicole Iturriaga. "How White Nationalists Mobilize Genetics: From Genetic Ancestry and Human Biodiversity to Counterscience and Metapolitics." *American Journal of Physical Anthropology* 175, no. 2 (2020): 387–98.

Panofsky, Aaron, and Joan Donovan. "Genetic Ancestry Testing among White Nationalists: From Identity Repair to Citizen Science." *Social Studies of Science* 49, no. 5 (2019): 653–81.

Papageorge, Nicholas W., and Kevin Thom. "Genes, Education, and Labor Market Outcomes: Evidence from the Health and Retirement Study." *National Bureau of Economic Research Working Paper* (2018). https://nber.org/papers/w25114.

Parker, Kim, Juliana Menasce Horowitz, and Anna Brown. "Americans' Complex Views on Gender Identity and Transgender Issues." Pew Research Center, June 28, 2022. https://pewresearch.org/social-trends/2022/06/28/americans-complex-views-on-gender-identity-and-transgender-issues.

Pastore, Nicholas. "The Army Intelligence Tests and Walter Lippmann." *Journal of the History of the Behavioral Sciences* 14, no. 4 (1978): 316–27.

Pearson, Susan J. *The Birth Certificate: An American History.* Chapel Hill: University of North Carolina Press, 2022.

Perry, Samuel L. *Growing God's Family: The Global Orphan Care Movement and the Limits of Evangelical Activism.* New York: NYU Press, 2017.

Perry, Samuel L., and Andrew Whitehead. "Christian Nationalism, Racial Separatism, and Family Formation: Attitudes toward Transracial Adoption as a Test Case." *Race and Social Problems* 7, no. 2 (2015): 123–34.

Philippi, Cristian Larroulet. "On Measurement Scales: Neither Ordinal nor Interval?" *Philosophy of Science* 88 (2021): 929–39.

Pigati, Jeffrey S., Kathleen B. Springer, Jeffrey S. Honke, David Wahl, Marie R. Champagne, Susan R. H. Zimmerman, Harrison J. Gray, Vincent L. Santucci, Daniel Odess, David Bustos, and Matthew R. Bennett. "Independent Age Estimates Resolve the Controversy of Ancient Human Footprints at White Sands." *Science* 382, no. 6666 (2023): 73–75.

Piper, John. "Red, Yellow, Black, and White—Could Every Race Come from Adam, Eve, and Noah?" *Desiring God*, August 26, 2016. https://desiringgod.org/interviews/red-yellow-black-and-white-could-every-race-come-from-adam-eve-and-noah.

———. "Structural Racism: The Child of Structural Pride." *Desiring God*, November 15, 2016. https://desiringgod.org/articles/structural-racism.

Plato. *The Republic.* New York: Modern Library, 1960.

Plomin, Robert. *Blueprint: How DNA Makes Us Who We Are.* Cambridge, MA: MIT Press, 2018.

Plomin, Robert, and C. S. Bergeman. "The Nature of Nurture: Genetic Influence on 'Environmental' Measures." *Behavioral and Brain Sciences* 14, no. 3 (1991): 414–27.

Plomin, Robert, and Richard Rende. "Human Behavioral Genetics." *Annual Review of Psychology* 42, no. 1 (1991): 161–90.

Plotz, David. *The Genius Factory: The Curious History of the Nobel Prize Sperm Bank.* New York: Penguin Random House, 2005.

Popejoy, Alice B., K. R. Crooks, S. M. Fullerton, L. A. Hindorff, G. W. Hooker, B. A. Koenig, N. Pino, E. M. Ramos, D. I. Ritter, H. Wand, M. W. Wright, M. Yudell, J. Y. Zou, S. E. Plon, C. D. Bustamante, K. E. Ormond, and the Clinical Genome Resource (ClinGen) Ancestry and Diversity Working Group. "Clinical Genetics Lacks Standard Definitions and Protocols for the Collection and Use of Diversity Measures." *American Journal of Human Genetics* 107, no. 1 (2020): 72–82.

Popejoy, Alice B., and Stephanie M. Fullerton. "Genomics Is Failing on Diversity." *Nature* 538, no. 7624 (2016): 161–64.

Popenoe, Paul, and Roswell Hill Johnson. *Applied Eugenics.* New York: Macmillan, 1918.

Popkin, Richard H. "Pre-Adamism in 19th Century American Thought: 'Speculative Biology' and Racism." *Philosophia* 8, no. 2–3 (1978): 205–39.

Pray, Leslie A. "Discovery of DNA Structure and Function: Watson and Crick." *Scitable* 1, no. 1 (2008): 100.

Prewitt, Kenneth. *What Is "Your" Race? The Census and Our Flawed Efforts to Classify Americans.* Princeton, NJ: Princeton University Press, 2013.

Price, Alkes, Nick Patterson, Fuli Yu, David R. Cox, Alicja Waliszewska, Gavin J. McDonald, Arti Tandon, Christine Schirmer, Julie Neubauer, Gabriel Bedoya, Constanza Duque, Alberto Villegas, Maria Cátira Bortolini, Francisco M. Salzano, Carla Gallo, Guido Mazzotti, Marcela Tello-Ruíz, Laura Riba, Carlos A. Aguilar-Salinas, Samuel Canizales-Quinteros, Marta Menjivar, William Klitz, Brian Henderson, Christopher A. Haiman, Cheryl Winkler, Teresa Tusie-Luna, Andrés Ruiz-Linares, and David Reich. "A Genome Wide Admixture Map for Latino Populations." *American Journal of Human Genetics* 80, no. 6 (2007): 1024–26.

Public Religion Research Institute. "Religion and Congregations in a Time of Social and Political Upheaval." May 16, 2023. https://prri.org/research/religion-and-congregations -in-a-time-of-social-and-political-upheaval.

———. "Summer Unrest over Racial Injustice Moves the Country, but Not Republicans or White Evangelicals." August 21, 2020. https://prri.org/press-release/summer-unrest-over -racial-injustice-moves-the-country-but-not-republicans-or-white-evangelicals.

Putnam, Hilary. *The Threefold Cord: Mind, Body, and World.* New York: Columbia University Press, 1999.

Quiroga, Seline Szupinski. "Blood Is Thicker than Water: Policing Donor Insemination and the Reproduction of Whiteness." *Hypatia* 22, no. 2 (2007): 143–61.

Raff, Jennifer. *Origin: A Genetic Story of the Americas.* New York: Twelve Books, 2022.

Rafter, N. "Earnest A. Hooton and the Biological Tradition in American Criminology." *Criminology* 42 (2004): 735–72.

Reardon, Jennifer. *Race to the Finish: Identity and Governance in an Age of Genomics.* Princeton, NJ: Princeton University Press, 2005.

Reardon, Jenny, and Kim TallBear. "'Your DNA Is Our History': Genomics, Anthropology, and the Construction of Whiteness as Property." *Current Anthropology* 53, supplement 5 (2012): S233–45.

Reich, David. "How to Talk about 'Race' and Genetics." *New York Times*, March 30, 2018. https://nytimes.com/2018/03/30/opinion/race-genetics.html.

———. "Race in the Age of Modern Genetics." *New York Times*, Sunday Review, March 23, 2018. https://nytimes.com/2018/03/23/opinion/sunday/genetics-race.html.

———. *Who We Are and How We Got Here: Ancient DNA and the New Science of the Human Past*. New York: Pantheon Books, 2018.

Reich, David, Nick Patterson, Desmond Campbell, Arti Tandon, Stéphanie Mazieres, Nicolas Ray, Maria V. Parra, Winston Rojas, Constanza Duque, Natalia Mesa, et al. "Reconstructing Native American Population History." *Nature* 488, no. 7411 (2012): 370–75.

Relethford, John H. "Biological Anthropology, Population Genetics, and Race." In *The Oxford Handbook of Philosophy and Race*, ed. Naomi Zack, 160–69. New York: Oxford University Press, 2017.

Rhodes, M. "How Two Intuitive Theories Shape the Development of Social Categorization." *Child Development Perspectives* 7 (2013): 12–16.

Rich, Camille Gear. "Contracting Our Way to Inequality: Race, Reproductive Freedom, and the Quest for the Perfect Child." *Minnesota Law Review* 104 (2019): 2375–469.

Rietveld, Cornelius A., Sarah E. Medland, Jaime Derringer, Jian Yang, Tõnu Esko, Nicolas W. Martin, Harm-Jan Westra, Konstantin Shakhabazov, Abdel Abdellaoui, Arpana Agrawal, et al. "GWAS of 126,559 Individuals Identifies Genetic Variants Associated with Educational Attainment." *Science* 340, no. 6139 (2013): 1467–71.

Rife, David C. "Populations of Hybrid Origin as Source Material for the Detection of Linkage." *American Journal of Human Genetics* 6, no. 1 (1954): 26–33.

Roberts, Dorothy. "Debating the Cause of Health Disparities: Implications for Bioethics and Racial Equality." *Cambridge Quarterly of Healthcare Ethics* 21 (2012): 332–41.

———. *Fatal Invention: How Science, Politics, and Big Business Re-Create Race in the Twenty-First Century*. New York: The New Press, 2011.

———. *Killing the Black Body: Race, Reproduction, and the Meaning of Liberty*. New York: Vintage Books, 1997.

———. *Shattered Bonds: The Color of Child Welfare*. London: Hachette UK, 2009.

———. "What's Race Got to Do with Medicine?" *Ted Radio Hour*, February 10, 2017. https://npr.org/transcripts/514150399.

Rodriguez, Clara E. *Changing Race: Latinos, the Census, and the History of Ethnicity in the United States*. New York: NYU Press, 2000.

Rogers, Adam. *Full Spectrum: How the Science of Color Made Us Modern*. New York: Houghton Mifflin Harcourt, 2021.

Roman, Youssef. "Race and Precisions Medicine: Is It Time for an Upgrade?" *Pharmacogenomics Journal* 19 (2019): 1–4.

Root, Michael. "Race in the Biomedical Sciences." In *The Oxford Handbook of Philosophy and Race*, ed. Naomi Zack, 463–73. New York: Oxford University Press, 2017.

———. "The Use of Race in Medicine as a Proxy for Genetic Differences." *Philosophy of Science* 70, no 5 (2003): 1173–83.

Rosen, Christine. *Preaching Eugenics: Religious Leaders and the American Eugenics Movement.* New York: Oxford University Press, 2004.

Rosenberg, Noah A., Saurabh Mahajan, Sohini Ramachandran, Chengfeng Zhao, Jonathan K. Pritchard, and Marcus W. Feldman. "Clines, Clusters, and the Effect of Study Design on the Inference of Human Population Structure." *PLOS Genetics* 1, no. 6 (2005): e70.

Rosenberg, Noah A., Jonathan K. Pritchard, James L. Weber, Howard M. Cann, Kenneth K. Kidd, Lev A. Zhivotovsky, and Marcus W. Feldman. "Genetic Structure of Human Populations." *Science* 298, no. 5602 (2002): 2381–85.

Ross, L.N. "Multiple Realizability from a Causal Perspective." *Philosophy of Science* 87 (2020): 640–62.

Roth, Wendy D. "The Multiple Dimensions of Race." *Ethnic and Racial Studies* 39, no. 8 (2016): 1310–38.

Rothman, Barbara Katz. *The Book of Life: A Personal and Ethical Guide to Race, Normality and the Human Gene Study.* Boston: Beacon Press, 2001.

Roustam, Raza. *Napoleon's Mameluke: The Memoirs of Roustam Raeza.* New York: Enigma Books, 2015.

Rowe, David C. *The Limits of Family Influence: Genes, Experience, and Behavior.* New York: Guilford Press, 1994.

Ruiz-Linares, Andrés, Kaustubh Adhikari, Victor Acuña-Alonzo, Mirsha Quinto-Sanchez, Claudia Jaramillo, William Arias, Macarena Fuentes, et al. "Admixture in Latin America: Geographic Structure, Phenotypic Diversity and Self-Perception of Ancestry Based on 7,342 Individuals." *PLOS Genetics* 10, no. 9 (2014): e1004572.

Russell, Camisha A. *The Assisted Reproduction of Race.* Bloomington: Indiana University Press, 2018.

Saini, Angela. *Superior: The Return of Race Science.* Boston: Beacon Press, 2019.

Salzano, Francisco M. "Interethnic Variability and Admixture in Latin America—Social Implications." *Revista Biología Tropical* 52, no. 3 (2004): 405–15.

Sans, Mónica. "Admixture Studies in Latin America: From the 20th to the 21st Century." *Human Biology* 72, no. 1 (2000): 155–77.

Sapolsky, Robert M. "A Height Gene? One for Smarts? Don't Bet on It." *Wall Street Journal*, January 31, 2014. https://ssgac.org/documents/New-Evidence-That-Genes-Play-a-Minimal-Role-in-Many-Traits-WSJ.pdf.

Satel, Sally. "I Am a Racially Profiling Doctor." *New York Times*, May 5, 2002. https://nytimes.com/2002/05/05/magazine/i-am-a-racially-profiling-doctor.html.

Scales-Trent, Judy. "Racial Purity Laws in the United States and Nazi Germany: The Targeting Process." *Human Rights Quarterly* 259 (2001): 260–307.

Scarr, Sandra. "On Arthur Jensen's Integrity." *Intelligence* 26, no. 3 (1998): 227–32.

Schechter, M.S., N. Sabater-Anaya, G. Oster, D. Weycker, H. Wu, E. Arteaga-Solis, S. Bagal, L.J. McGarry, K. Van Brunt, and J.M. Geiger. "Impact of Elexacaftor/Tezacaftor/Ivacaftor on Healthcare Resource Utilization and Associated Costs among People with Cystic Fibrosis in the U.S.: A Retrospective Claims Analysis." *Pulmonary Therapy* 9, no. 4 (2023): 479–98.

Scheib, Joanna. "The Psychology of Female Choice in the Context of Donor Insemination." In *Darwinian Feminism and Human Affairs*, 489–504. New York: Springer Science and Business Media, 1997.

Scheitle, Christopher P., and Katie E. Corcoran. "COVID-19 Skepticism in Relation to Other Forms of Science Skepticism." *Socius* 7 (2021). https://doi.org/10.1177/23780231211049841.

Science News staff. "Some Past *Science News* Coverage Was Racist and Sexist. We're Deeply Sorry." *Science News*, March 24, 2022. https://www.sciencenews.org/article/past-science-news-coverage-racism-sexism.

Sey, Samuel. "Murders That Won't Go Viral: The Quiet Injustice Too Few Protest." *Desiring God*, October 30, 2020. https://desiringgod.org/articles/murders-that-wont-go-viral.

Shapiro, Beth, Axel Barlow, Peter D. Heintzman, Michael Hofreiter, Johanna L. A. Paijmans, and André E. R. Soares, eds. *Ancient DNA: Methods and Protocols*. 2nd ed. New York: Human Press, [2012] 2019.

Sheridan, Cormac. "The World's First CRISPR Therapy Is Approved: Who Will Receive It?" *Nature Biotechnology* 42, no. 1 (2024): 3–4.

Shiao, Jiannbin Lee, Thomas Bode, Amber Beyer, and Daniel Selvig. "The Genomic Challenge to the Social Construction of Race." *Sociological Theory* 30, no. 2 (2012): 67–88.

Shurkin, Joel N. *Broken Genius: The Rise and Fall of William Shockley, Creator of the Electronic Age*. New York: Palgrave Macmillan, 2006.

———. *Terman's Kids: The Groundbreaking Study of How the Gifted Grow Up*. Boston: Little, Brown, 1992.

Skoglund, Pontus, Swapan Mallick, Maria Cátira Bortolini, Niru Chennagiri, Tábita Hünemeier, Maria Luiza Petzl-Erler, Francisco Mauro Salzano, Nick Patterson, and David Reich. "Genetic Evidence for Two Founding Populations of the Americas." *Nature* 525, no. 7567 (2015): 104–8.

Skoglund, Pontus, and David Reich. "A Genomic View of the Peopling of the Americas." *Current Opinion in Genetics and Development* 41 (2016): 27–35.

Skotte, Line, Thorfinn Sand Korneliussen, and Anders Albrechtsen. "Estimating Individual Admixture Proportions from Next Generation Sequencing Data." *Genetics* 195, no. 3 (2013): 693–702.

Smart, Andrew, Deborah A. Bolnick, and Richard Tutton. "Health and Genetic Ancestry Testing: Time to Bridge the Gap." *BMC Medical Genomics* 10, no. 3 (2017). https://link.springer.com/article/10.1186/s12920-016-0240-3.

Smart, Andrew, Richard Tutton, Paul Martin, George T. H. Ellison, and Richard Ashcroft. "The Standardisation of Race and Ethnicity in Biomedical Science Editorials and UK Biobanks." *Social Studies of Science* 38, no. 3 (2007): 407–23.

Smedley, Audrey. *Race in North America: Origin and Evolution of a Worldview*. New York: Routledge, 2018.

Smith, Greg. "More White Americans Adopted Than Shed Evangelical Label During Trump Presidency, Especially His Supporters." Pew Research Center, September 15, 2021. https://pewresearch.org/fact-tank/2021/09/15/more-white-americans-adopted-than-shed-evangelical-label-during-trump-presidency-especially-his-supporters.

Smith, J. David, and Michael L. Wehmeyer. "Who Was Deborah Kallikak?" *Intellectual and Developmental Disabilities* 50, no. 2 (2012): 169–78.

Smith, Michael W., and Stephen J. O'Brien. "Mapping by Admixture Linkage Disequilibrium: Advances, Limitations and Guideline." *Nature Reviews Genetics* 6, no. 8 (2005): 623–32.

Soni, Sheetal, and Julian Savulescu. "Polygenic Embryo Screening: Ethical and Legal Considerations." Hastings Center, October 20, 2021. https://www.thehastingscenter.org/polygenic-embryos-screening-ethical-and-legal-considerations.

Spar, Debora. *The Baby Business: How Markets Are Changing the Future of Birth*. Cambridge, MA: Harvard Business School Press, 2006.

Spiro, Jonathan. *Defending the Master Race: Conservation, Eugenics, and the Legacy of Madison Grant*. Lebanon, NH: University Press of New England, 2009.

Stein, Rob. "FDA Advisers See No Roadblock for Gene-Editing Treatment for Sickle Cell Disease." NPR, October 31, 2023. https://npr.org/sections/health-shots/2023/10/31/1208041252/a-landmark-gene-editing-treatment-for-sickle-cell-disease-moves-closer-to-reality.

Stephens, J. Claiborne, David Briscoe, and Stephen J. O'Brien. "Mapping by Admixture Linkage Disequilibrium in Human Populations: Limits and Guidelines." *American Journal of Human Genetics* 55, no. 4 (1994): 809–24.

Stern, Alexandra Minna. *Eugenic Nation: The Faults and Frontiers of Better Breeding in America*. Berkeley: University of California Press, 2005.

———. "From 'Race Suicide' to 'White Extinction': White Nationalism, Nativism, and Eugenics over the Past Century." *Journal of American History* 109, no. 2 (2022): 348–61.

———. *Telling Genes: The Story of Genetic Counseling in America*. Baltimore: Johns Hopkins University Press, 2012.

Stoddard, Lothrop. *The Revolt against Civilization: The Menace of the Under Man*. New York: Charles Scribner's Sons, 1922.

Suarez-Kurtz, Guilherme. *Pharmacogenomics in Admixed Populations*. Austin: Landes Bioscience, 2007.

Subica, Andrew M. "CRISPR in Public Health: The Health Equity Implications and Role of Community in Gene-Editing Research and Applications." *American Journal of Public Health* 113, no. 8 (2023): 874–82.

Suits, Steve. *Overturning Brown: The Segregationist Legacy of the Modern School Choice Movement*. Montgomery: New South Books, 2020.

Sunder Rajan, Kaushik. *Biocapital: The Constitution of Postgenomic Life*. Durham, NC: Duke University Press, 2006.

Sussman, Robert Wald. *The Myth of Race: The Troubling Persistence of an Unscientific Idea*. Cambridge, MA: Harvard University Press, 2014.

Szentpetery, Sylvia E., Erick Forno, Glorisa Canino, and Juan C. Celedón. "Asthma in Puerto Ricans: Lessons from a High-Risk Population." *Journal of Allergy and Clinical Immunology* 138, no. 6 (2016): 1556–58.

TallBear, Kim. *Native American DNA: Tribal Belonging and the False Promise of Genetic Science*. Minneapolis: University of Minnesota Press, 2013.

Tamm, Erika, Toomas Kivisild, Maere Reidla, Mait Metspalu, David Glenn smith, Connie J. Mulligan, Claudio M. Bravi, Olga Rickards, Cristina Martinez-Labarga, Elsa K. Khusnutdinova, et al. "Beringian Standstill and Spread of Native American Founders." *PLOS One* 2, no. 9 (2007): e829.

Target. "Ivory Original Bar Soap." https://www.target.com/p/ivory-original-bar-soap-10pk-3-17oz-each-it-floats/-/A-13951811.

Taylor, Howard F. *The IQ Game: A Methodological Inquiry into the Heredity-Environment Controversy*. New Brunswick, NJ: Rutgers University Press, 1980.

Terman, Lewis Madison. *Genetic Studies of Genius*. Vol. 1, *Mental and Physical Traits of a Thousand Gifted Children*. Stanford, CA: Stanford University Press, 1926.

———. "The Influence of Nature and Nurture upon Intelligence Scores: An Evaluation of the Evidence in Part I of the 1928 Yearbook of the National Society for the Study of Education." *Journal of Educational Psychology* 19, no. 6 (1928): 362–73.

———. *The Measurement of Intelligence*. Boston: Houghton Mifflin, 1916.

Thompson, Warren S. "Race Suicide in the United States." *Scientific Monthly* 5, no. 1 (1917): 22–35.

1000 Genomes Project Consortium. "A Global Reference for Human Genetic Variation." *Nature* 526, no. 7571 (2015): 68–74.

Tielbeek, Jorim J., Ada Johansson, Tinca J. C. Polderman, Marja-Ritta Rautiainen, Philip Jansen, Michelle Taylor, Xiaoran Tong, Qing Lu, Alexandra S. Burt, Henning Tiemeier, et al. "Genome-Wide Association Studies of a Broad Spectrum of Antisocial Behavior." *JAMA Psychiatry* 74 (2017): 1242–50.

Tisby, Jemar. *The Color of Compromise: The Truth about the American Church's Complicity in Racism*. Grand Rapids, MI: Zondervan Reflective, 2019.

Tucci, Serena, and Joshua M. Akey. "The Long Walk to African Genomics." *Genome Biology* 20 (2019): 1–3.

Tucker, William H. *The Funding of Scientific Racism: Wickliffe Draper and the Pioneer Fund*. Champaign: University of Illinois Press, 2002.

Turkheimer, Eric. "Peter Visscher on the Genomics of Complex Human Traits." May 23, 2022. https://turkheimer.com/peter-visscher-on-the-genomics-of-complex-human-traits.

Turkheimer, Eric, and Irving I. Gottesman. "Is h^2 = 0 a Null Hypothesis Anymore?" *Behavioral and Brain Sciences* 14, no. 3 (1991): 410–11.

Turley, Patrick, Michelle N. Meyer, Nancy Wang, David Cesarini, Evelynn Hammonds, Alicia R. Martin, Benjamin M. Neale, Heidi L. Rehm, Louise Wilkins-Haug, Daniel J. Benjamin, Steven Hyman, David Laibson, and Peter M. Visscher. "Problems with Using Polygenic Scores to Select Embryos." *New England Journal of Medicine* 385 (2021): 78–86.

Underhill, Megan R. "'Diversity Is Important to Me': White Parents and Exposure-to-Diversity Parenting Practices." *Sociology of Race and Ethnicity* 5, no. 4 (2019): 486–99.

UNESCO. *Four Statements on the Race Question*. Paris: United Nations Educational, Scientific and Cultural Organization, 1969.

United Nations. "Convention on the Prevention and Punishment of the Crime of Genocide." 1948.

US Birth Certificates. "How Is Race Determined on Birth Certificates?" https://www.usbirthcertificates.com/articles/race-on-birth-certificates.

US Census Bureau. "Measuring Race and Ethnicity across the Decades, 1790–2010." https://www.census.gov/data-tools/demo-race/MREAD_1790_2010.html.

US Department of Health and Human Services, Office of Minority Health. "Asthma and Hispanic Americans." https://minorityhealth.hhs.gov/asthma-and-hispanic-americans.

US Food and Drug Administration. "What You Should Know—Reproductive Tissue Donation." November 5, 2010. https://www.fda.gov/vaccines-blood-biologics/safety-availability-biologics/what-you-should-know-reproductive-tissue-donation.

Valdez, Natali, and Daisy Deomampo. "Centering Race and Racism in Reproduction." *Medical Anthropology* 38, no. 7 (2019): 551–59.

Velleman, P., and L. Wilkinson. "Nominal, Ordinal, Interval, and Ratio Typologies are Misleading." In *Trends and Perspectives in Empirical Social Research*, ed. Ingwer Borg and Peter P. Mohler, 161–77. Berlin: De Gruyter, 2011.

Vienne, J. M. "Locke on Real Essence and Internal Constitution." *Proceedings of the Aristotelian Society* 93 (1993): 139–53.

Visscher, M. O., S. A. Burkes, D. M. Adams, A. M. Hammill, and R. R. Wickett. "Infant Skin Maturation: Preliminary Outcomes for Color and Biomechanical Properties." *Skin Research Technology* 23 no. 4 (November 2017): 545–51.

Vitti, Joseph. "Opinion: Big Data Scientists Must Be Ethicists Too." *Broad Minded*, August 29, 2019. https://www.broadinstitute.org/blog/opinion-big-data-scientists-must-be-ethicists-too.

Voosen, Paul. "There Is No Gene for Finishing College." *Chronicle of Higher Education*, May 30, 2013. https://www.chronicle.com/blogs/percolator/there-is-no-gene-for-finishing-college.

Vora, Kalindi. "Indian Transnational Surrogacy and the Commodification of Vital Energy." *Subjectivity* 28, no. 1 (2009): 266–78.

Vyas, A., Leo G. Eisenstein, and David S. Jones. "Hidden in Plain Sight—Reconsidering the Use of Race Correction in Clinical Algorithms." *New England Journal of Medicine* 383 (2020): 874–82.

Wade, Nicholas. "Earliest Americans Arrived in Waves, DNA Study Finds." *New York Times*, July 12, 2012.

———. *A Troublesome Inheritance: Genes, Race and Human History.* New York: Penguin, 2014.

Wailoo, Keith, Alondra Nelson, and Catherine Lee, eds. *Genetics and the Unsettled Past: The Collision of DNA, Race, and History.* New Brunswick, NJ: Rutgers University Press, 2012.

Wang, Amy B. "Nivea's 'White is Purity' Ad Campaign Did Not Go Well." *Los Angeles Times*, April 5, 2017, https://latimes.com/business/la-fi-nivea-white-20170405-story.html.

Wang, Sijia, Nicolas Ray, Winston Rojas, Maria V. Parra, Gabriel Bedoya, Carla Gallo, Giovanni Poletti, et al. "Geographic Patterns of Genome Admixture in Latin American Mestizos." *PLOS Genetics* 4, no. 3 (2008): 1–9.

Ware, Lawrence. "If Your Pastor Says 'Racism Isn't a Skin Problem, It's a Sin Problem' You Need to Find Another Church." *The Root*, August 16, 2016. https://theroot.com/if-your-pastor-says-racism-isn-t-a-skin-problem-it-s-1822522858.

Warren, Rick. "How to Overcome the Sin of Prejudice." June 15, 2020. https://pastorrick.com/how-to-overcome-the-sin-of-prejudice.

———. "You're Called to Belong to the Church." August 22, 2019. https://pastorrick.com/youre-called-to-belong-to-the-church.

Wedgworth, Steven. "The Science of Male and Female: What God Teaches through Nature." *Desiring God*, September 11, 2020, https://desiringgod.org/articles/the-science-of-male-and-female.

Wedow, Robbee, Daphne O. Martschenko, and Sam Trejo. "Scientists Must Consider the Risk of Racist Misappropriation of Research." *Scientific American*, May 26, 2022. https://scientificamerican.com/article/scientists-must-consider-the-risk-of-racist-misappropriation-of-research.

Wei, Wei. "Queering the Rise of China: Gay Parenthood, Transnational ARTs, and Dislocated Reproductive Rights." *Feminist Studies* 47, no. 2 (2021): 312–40.

Weikel, Blair W., Susanne Klawetter, Stephanie L. Bourque, Kathleen E. Hannan, Kristi Roybal, Modi Soondarotok, Marie St. Pierre, Yarden S. Fraiman, and Sunah S. Hwang. "Defining an Infant's Race and Ethnicity: A Systematic Review." *Pediatrics* 151, no. 1 (January 1, 2023): e2022058756.

Wells, Spencer. *Deep Ancestry: Inside the Genographic Project*. Washington, DC: National Geographic Society, 2006.

———. *The Journey of Man: A Genetic Odyssey*. London: Allen Lane, 2002.

Wesseldijk, L. W., et al. "Using a Polygenic Score in a Family Design to Understand Genetic Influences on Musicality." *Scientific Reports* 12 (2022): 14658.

Wilde, Melissa J. *Birth Control Battles: How Race and Class Divided American Religion*. Oakland: University of California Press, 2019.

———. "Complex Religion: Interrogating Assumptions of Independence in the Study of Religion." *Sociology of Religion* 79, no. 3 (2018): 287–98.

Williams, Patricia J. "Babies, Bodies and Buyers." *Columbia Journal of Gender and Law* 33 (2016): 11–24.

Wilson, P. K. "Harry Laughlin's Eugenic Crusade to Control the 'Socially Inadequate' in Progressive Era America." *Patterns of Prejudice* 36, no. 1 (2022): 49–67.

Winkler, Cheryl A., George W. Nelson, and Michael W. Smith. "Admixture Mapping Comes of Age." *Annual Review of Genomics and Human Genetics* 11 (2010): 65–89.

Wojcicki, Anne. "Letter to Investors," 2007. https://assets.bwbx.io/images/users/iqjWHBFdfxIU/i_JWmL6tfMhA/v2/1650x2200.jpg.

Wright, Sewall. "Correlation and Causation." *Journal of Agricultural Research* 20, no. 7 (1921): 557–85.

Wu, Chia-Ling. "Managing Multiple Masculinities in Donor Insemination: Doctors Configuring Infertile Men and Sperm Donors in Taiwan." *Sociology of Health and Illness* 33, no. 1 (2010): 96–113.

Yerkes, Robert M. "Eugenic Bearing of Measurements of Intelligence." *Eugenics Review* 24, no. 4 (1923): 225–45.

Young, Toby. "The Fall of the Meritocracy." *Quadrant Online*, September 7, 2015. https://quadrant.org.au/magazine/2015/09/fall-meritocracy.

Yuan, Christopher. "He Made Them Male and Female: Sex, Gender, and the Image of God." *Desiring God*, December 14, 2019, https://desiringgod.org/articles/he-made-them-male-and-female.

Zembylas, Michalinos. "Affect, Race, and White Discomfort in Schooling: Decolonial Strategies for 'Pedagogies of Discomfort.'" *Ethics and Education* 13, no. 1 (2018): 86–104.

Zenderland, Leila. *Measuring Minds: Henry Herbert Goddard and the Origins of American Intelligence Testing*. New York: Cambridge University Press, 1998.

Zhang, Sarah. "The Fertility Doctor's Secret." *Atlantic*, April 2019. https://www.theatlantic.com/magazine/archive/2019/04/fertility-doctor-donald-cline-secret-children/583249.

Zimmer, Carl. *She Has Her Mother's Laugh: The Powers, Perversions, and Potential of Heredity*. New York: Dutton, 2018.

CONTRIBUTORS

CARLOS ANDRÉS BARRAGÁN, PhD, Department of Science and Technology Studies, UC Davis, cabarragan@ucdavis.edu

MARK FEDYK, PhD, UC Davis School of Medicine, mfedyk@ucdavis.edu

CHERIE GINWALLA, MD, Department of Pediatrics, UC Davis School of Medicine, cfginwalla@ucdavis.edu

JAMES GRIESEMER, PhD, Departments of Philosophy and Science and Technology Studies, UC Davis, jrgriesemer@ucdavis.edu

LISA C. IKEMOTO, JD, LLM, UC Davis School of Law, lcikemoto@ucdavis.edu

EMILY KLANCHER MERCHANT, PhD, Department of Science and Technology Studies, UC Davis, ekmerchant@ucdavis.edu

MEAGHAN O'KEEFE, PhD, Department of Religious Studies, UC Davis, mmokeefe@ucdavis.edu

ALICE B. POPEJOY, PhD, Department of Public Health Sciences (Epidemiology) and UCD Comprehensive Cancer Center, UC Davis Health, abpopejoy@ucdavis.edu

TINA RULLI, PhD, Department of Philosophy, UC Davis, trulli@ucdavis.edu

SIVAN YAIR, PhD, Department of Evolution and Ecology, UC Davis, ssyair@ucdavis.edu

INDEX

Founded in 1893,
UNIVERSITY OF CALIFORNIA PRESS
publishes bold, progressive books and journals
on topics in the arts, humanities, social sciences,
and natural sciences—with a focus on social
justice issues—that inspire thought and action
among readers worldwide.

The UC PRESS FOUNDATION
raises funds to uphold the press's vital role
as an independent, nonprofit publisher, and
receives philanthropic support from a wide
range of individuals and institutions—and from
committed readers like you. To learn more, visit
ucpress.edu/supportus.